Proceedings of SPIE—The International Society for Optical Engineering

Volume 685

Infrared Technology XII

Irving J. Spiro, Richard A. Mollicone
Chairs/Editors

Sponsored by
SPIE—The International Society for Optical Engineering

Cooperating Organizations
Institute of Optics/University of Rochester
Jet Propulsion Laboratory/California Institute of Technology
Optical Sciences Center/University of Arizona
NASA/Ames Research Center
Center for Applied Optics/University of Alabama in Huntsville
Center for Electro-Optics/University of Dayton

19–20 August 1986
San Diego, California

Published by
SPIE—The International Society for Optical Engineering
P.O. Box 10, Bellingham, Washington 98227-0010 USA
Telephone 206/676-3290 (Pacific Time) • Telex 46-7053

SPIE (The Society of Photo-Optical Instrumentation Engineers) is a nonprofit society dedicated to advancing engineering and scientific applications of optical, electro-optical, and optoelectronic instrumentation, systems, and technology.

The papers appearing in this book comprise the proceedings of the meeting mentioned on the cover and title page. They reflect the authors' opinions and are published as presented and without change, in the interests of timely dissemination. Their inclusion in this publication does not necessarily constitute endorsement by the editors or by SPIE.

Please use the following format to cite material from this book:
Author(s), "Title of Paper," *Infrared Technology XII,* Irving J. Spiro, Richard A. Mollicone, Editors, Proc. SPIE 685, page numbers (1986).

Library of Congress Catalog Card No. 85-51815
ISBN 0-89252-720-X
ISSN 8755-180

Printed in the United States of America.

INFRARED TECHNOLOGY XII

Volume 685

Contents

INFRARED TECHNOLOGY XII

Volume 685

Conference Committee

Chairs
Irving J. Spiro
The Aerospace Corporation

Richard A. Mollicone
Analytic Decisions Incorporated

Session Chairs
Session 1—Infrared Imaging
Irving J. Spiro, The Aerospace Corporation

Session 2—Simulation, Modeling, and Testing
Randall Murphy, Air Force Geophysics Laboratory

Session 3—Infrared in the United Kingdom
Douglas E. Burgess, Peter N. J. Dennis, Royal Signals
and Radar Establishment, United Kingdom

Session 4—Infrared Applications
Richard A. Mollicone, Analytic Decisions Incorporated

Conference 685, *Infrared Technology XII,* was part of a three-conference program on Infrared and Ultraviolet Technologies held at SPIE's 30th Annual International Technical Symposium on Optical and Optoelectronic Applied Sciences and Engineering. The other conferences were

Conference 686, *Infrared Detectors, Sensors, and Focal Plane Arrays*
Conference 687, *Ultraviolet Technology*

Program Chair: **Irving J. Spiro,** The Aerospace Corporation

INTRODUCTION

The twelfth annual conference on Infrared Technology was held in San Diego, California, on August 19 and 20, 1986, and was comprised of four sessions.

Session 1: Infrared Imaging. Holst of Martin Marietta Aerospace opened the session with a paper on a semiautomated minimum resolvable temperature (MRT) technique. While methods for automation of noise equivalent temperature difference (NETD) and signal transfer function (SITF) exist, methods to automate MRT have had limited success. This paper proposes a semiautomated method of obtaining MRT that accounts for the differences between observers and their learning curves.

Duvall of the Jet Propulsion Laboratory describes the VIMS, an imaging spectrometer for investigation of Mars. The instrument has two linear arrays—a 64-element silicon and a 256-element InSb. The unit will map Mars in the 0.3 to 4.3 μm region at a resolution of about 1 km. A V-groove radiator cools the detectors to 88 K.

Miller of Unilever Research Port Sunlight Laboratory (UK) discusses thermal wave imaging as a nondestructive technique for a wide range of practical applications. Optically induced heat fluxes are used to probe the physical and chemical properties of solid samples.

Full field stress analysis using the thermoelastic principle is described by Baker of Sira Ltd. (UK) and Oliver of Ometron Inc. The use of high sensitivity IR detectors coupled with fast electronic signal processing has led to the development of instrumentation capable of mapping dynamic stresses in structures; it is called stress pattern analysis by measurement of thermal emission—SPATE.

Aronson, Cenker, and Gilmartin of RCA Astro-Electronics describe a shuttle IR imaging experiment. A PtSi array was used aboard the January 1986 shuttle flight. Imagery shown included rivers, lakes, volcanoes, the earth limb, and the airglow layer. Cenker was the payload specialist aboard the Columbia and details not only the instrumentation but also its integration into the shuttle bay.

Li of Wuhan University (China) writes on signal processing of an IR imaging system.

Session 2: Simulation, Modeling, and Testing. In this session, Gary Wilson of Ball Aerospace Systems Division substituted as chair for Randall Murphy of the Air Force Geophysics Laboratory. In the first paper, Schott of Rochester Institute of Technology discusses the incorporation of angular emissivity effects in long wave IR image models. The effect of emissivity on simulated images and the effect of angle of view on emissivity are presented. An example of contrast is shown as the angle of viewing a ship is varied from 90° to 0°.

Bates of Sandia National Laboratories and Giles of New Mexico State University describe an IR scene composer for electronic vision applications. First, a radiance image is created from the Planck equation modified by responsivity and emissivity in the 8 to 14 μm region. From this, a temperature image is created. The paper includes examples of real, synthetic, and merged IR images. A library of models has been accomplished.

The maximum likelihood estimation of point source target amplitude and position in a mismatched detector environment is described by Rochester of Ball Aerospace Systems Division. Algorithms are detailed and hardware is discussed that is well matched to implement the algorithm. An array of sample times and an array of coefficients in the form of weights applied to TDI steps represent differences in responsivities and noise. This produces the maximum likelihood, which yields the maximum probability of detection. The hardware that performs these operations is capable of 512 FFTs in 300 microseconds.

Cogliandro and Castelli of I. A. M. Rinaldo Piaggio (Italy) present plume IR signature measurements and compare them with a theoretical model. The goal was to measure IR engine signatures of rotor craft in the 1 to 15 μm region. The project aimed to suppress IR radiation in order to elude IR missiles. The instrumentation, the measurement conditions, the actual measurements, and the comparison with theoretical predictions are shown.

continued

Beebe et al. of McDonnell Douglas Astronautics give a description of the focal plane/detector test and evaluation laboratory at McDonnell Douglas in Huntington Beach. Detector measurements include D^*, responsivity, R_oA, uniformity, noise, and efficiency. The same measurements plus linearity are made on focal plane arrays. One percent accuracy is claimed down to one pico amp. Two filter wheels select the wavelength to be measured.

Russell et al. of The Aerospace Corporation describe tests of IR arrays on the Kuiper Airborne Observatory. They demonstrate TDI on two IBC and the bulk array. A visible image on a TV screen is used as a view finder. Twelve specific stars, such as alpha Ori, in the 10.5 to 12.5 μm region were picked up. Several valuable lessons that will be useful on future programs are discussed.

Session 3: Infrared in the United Kingdom. Burgess of RSRE presented a ten-year progress report of the pyroelectric vidicon, co-authored by Nixon and Ritchie of English Electric Valve Company (UK). Dramatic advances have been made from the first IR sensitive TV camera tubes of ten years ago. A germanium window transmits IR radiation to a pyroelectric target. Variations of surface charges are read by a scanning electron beam. This produced high quality images with minimal camera modifications.

Sensors using linear detector arrays are described for a variety of applications by Liddicoat and Mansi of Plessey Electronic Systems Research Ltd. (UK). The paper, co-authored by Burgess and Manning of RSRE, describes sensors employing germanium lenses, a chopper to moderate the incoming radiation, and 64-element linear arrays. Picture quality is equal to, and in some cases exceeds, high quality monochrome TV systems.

A compact, low power focal plane of cadmium mercury telluride-silicon hybrid is described by Balingall and Blenkinsop of RSRE in conjunction with Baker and Parsons of Mullard Southampton Ltd. (UK). Two readout methods are presented, both of which will operate in the 3 to 5 or 8 to 10 μm bands at 77 K.

Dann et al. of Marconi Command and Control Systems Ltd. (UK) present a paper about real-time applications of staring arrays. Microscanning techniques have been demonstrated that allow for smaller, lighter optics and a wider range of applications. The system produces excellent quality images in the 3 to 5 and 8 to 12 μm wavebands using arrays of up to 64×64 elements.

Braim of RSRE and Cuthbertson of GEC Avionics Ltd. (UK) describe a dual waveband imaging radiometer, capable of simultaneous imagery in both IR wavebands. Simplicity of recording data, subsequent ease of analysis, and performance equal to state-of-the-art FLIRs are key benefits of this system.

Aspects of thermal imaging sensors configured to fit in a tubular structure compatible with a range of periscope designs are discussed by Runciman of Barr and Stroud Ltd. (UK). A polycrystalline germanium dome was chosen to withstand the chemical and biological effects of sea water as well as high pressures experienced at submarine operating depths. The modular design provides for flexibility of use and ease of maintenance and allows for evolutionary improvements.

Lettington of RSRE and Moore of Rank Pullin Controls (UK) discuss a thermal imager based on a novel coaxial scanning technique. This miniature imager was originally designed for RPV use but has gained wide acceptance in various military applications where size, weight, and cost are at a premium. A small high speed rotating polygon with reflective facets forms the basis of the afocal scanner.

McEwen of GEC Avionics Ltd. (UK) discusses a signal processing unit developed for 2D cadmium mercury telluride-silicon hybrid PV detector arrays. Severe fixed pattern noise is removed in the analog domain without the use of digital signal processing or analog to digital convertors. The resulting compact high speed IR imaging system has been prototyped for use in automatic weapon guidance and homing systems, particularly for the antiarmor role.

Session 4: Infrared Applications. The session begins with Bates of Sandia National Labs. and Giles of New Mexico State University describing a space-variant PSF model of a spirally scanning IR system. An IR scene has been generated and their paper is based on the scan produced on the ground if this system were dropped while rotating at a constant spin rate and a constant fall velocity. The PSF is attained by convolution for both 10 and 30 milliradian scenes.

Price of the Air Force Geophysics Laboratory writes about extended emission in the celestial background. The paper (presented by Paul Levan) describes measurements of zodiacal light made by the ZIP and IRAS payloads. They are found to be 10% reflecting in the visible and 90% emitting in the IR. Charts are shown depicting the signals plotted against wavelength and plotted against position away from the sun.

continued

Baker et al. from NASA/Langley Research Center describe the design and performance of the halogen occultation experiment (HALOE) remote sensor. Two measurement techniques are used to determine the concentration of various gases; four by the two-path method and four by standard radiometry. The goal is to observe the sun at dawn and dusk to determine the effect of certain gases on the ozone layer.

In a companion paper, Moore et al. from NASA/Langley, Systems and Applied Science Corporation, and the University of Arizona describe the calibration of the HALOE instruments. In addition to the radiometric instruments, the azimuth and elevation controls are shown. The methods of fine tracking and the means of compensating for the air masses are discussed.

Hanley et al. from OptiMetrics and the U.S. Army Atmospheric Sciences Laboratory present a method for long-path transmissometry. Four spectral bands are used: visible, 1.06 μm, 3 to 5 and 8 to 14 μm over a 10 km path. The methods of data acquisition along with the measurement results are presented.

Cha of the University of Illinois at Chicago explains the application of an IR gas filter correlation spectrometer for measurement of methanol concentrations in automobile exhausts. A nondispersive IR analyzer using a gas-filter correlation spectrometer (GFCS) is used to measure CO, NO_x, hydrocarbons, and oxygenated hydrocarbons. The method of determining methanol concentrations using the GFCS is discussed and correlation with gas chromatography for measuring the gases is presented.

Liptak et al. of the Technical University, Košice, Czechoslovakia, write about monitorization technology applied to metal cutting.

Irving J. Spiro
The Aerospace Corporation

Richard A. Mollicone
Analytic Decisions Incorporated

INFRARED TECHNOLOGY XII

Volume 685

Session 1

Infrared Imaging

Chair
Irving J. Spiro
The Aerospace Corporation

Semi-automated MRT Technique

Gerald C. Holst

Martin Marietta Aerospace
Mail Stop 1078, P. O. Box 5837
Orlando, Florida 32855

Abstract

Thermal imaging systems are characterized by the minimum resolvable temperature (MRT), signal transfer function (SITF) and noise equivalent temperature difference (NEDT). While the NEDT and SITF can be automated, methods to automate the MRT have had limited success. A semi-automated MRT technique is introduced which is faster and more accurate than the classical method. The new method also identifies an observer's variability, distractions and learning curve.

Introduction

The characterization of a thermal imaging system can be divided into two major areas: quantitative measurements and subjective measurements. Effective circuit and optical design are characterized by a variety of quantitative measures which include the noise equivalent temperature difference (NEDT), signal transfer function (SITF), video level balance on a channel to channel basis, responsivity balance on a channel to channel basis, uniformity and system dynamic range. Subjective evaluation of image quality include flicker, shading, focus, the minimum resolvable temperature (MRT) and the effects of objectionable fixed or moving pattern noise.

The NEDT, SITF and the MRT are perhaps the most widely accepted measures of system performance. The MRT is of importance because it is the eye-brain interpretation of system sensitivity when the target is immersed in noise and it is considered related to target recognition and identification. The MRT includes the effects of noise, resolution, temporal and spatial integration. While the NEDT and SITF can be obtained in a relatively short time, the MRT can be tedious. Quantitative measures (NEDT and SITF) can in principle be automated and therefore are appropriate measures to be obtained by automatic test equipment. In order to establish automated subjective measures, a quantitative criterion must be established which is relatable to subjective evaluation. Methods to automate the MRT have been tried in the past but have met with limited success.

Current (classical) methods of measuring the MRT generally requires several individuals. One individual (the observer) optimizes the display for his perception of good imagery. The observer then tells another individual to raise or lower the temperature source until the four bar target (7:1 aspect ratio) just appears or disappears. The observer calls out raise or lower the target temperature several times until he is at his threshold of just detecting the four bars. The experiment is repeated for a negative contrast target (cold target). The MRT is defined as (T1-T2)/2 where T1 is the detection temperature of the hot target and T2 is the detection temperature of the cold target. By averaging the two values together, any offset bias in the thermal source is eliminated. The experiment is repeated at other spatial frequencies and several observers participate. The MRT values from all the ovservers at each spatial frequency are averaged together to obtain a composite MRT curve.

Semi-automated MRT

A new rapid semi-automated MRT procedure is introduced which is faster and considered more accurate than the classical method. In this method, the "up-and-down" technique is used in which the observer tracks his own threshhold. This technique is often used in psychophysical research dealing with threshold stimuli. The primary advantage of the "up-and-down" method is that it automatically concentrates testing about the mean. As shown in Figure 1, the observer controls the temperature source by pressing a button as he observes the target. He decreases the temperature when he can just perceive the target and increases the temperature when the target just disappears. The temperature is monitored continuous and graphed on a strip chart recorder. Statistical analysis of the temperature excursions yields both the mean and an estimate of the standard deviation of the MRT[1]. However for our application, it is sufficient to just determine the average value.

Experimental Methods

Ideally, the temperature should be monitored continuously on a strip chart recorder. For our experiment, the temperature source was controlled by a digital computer operating with an IEEE-488 bus. The temperature was sensed every second and plotted. The time was chosen to condense the temperature recordings. With this particular set up, temperature extremes may be missed. However the methodology does not significantly affect the determination of the average value. Shown in Figure 2 are typical traces of the computer plot.

The average detection temperature in Figure 2a is T1 = 0.048°C (hot target) and in Figure 2b the average is T2 = -0.3°C (cold target) which yields an MRT = 0.174°C. The thermal offset of -0.126°C is a consequence of the particular AC coupled thermal imaging system used and the particular test conditions.

Since the method also provides a graphical representation of observer variability (standard deviation), inter- and intra-observer consistency can be rapidly determined. Figure 3 illustrates a typical trace in which the observer was distracted for a few seconds. The average detection temperature of T = -0.4°C can still be obtained from the data even with the distraction. Usually a "learning" curve is associated with any test. That is, the detection threshold is initially high but the observer quickly adjusts the temperature about his minimum threshold. While this learning curve is not evident in the classical method, it becomes obvious in the semi-automated method (Figure 4).

Discussion

It should be recognized that with the semi-automated method the temperature source may not have adequate time to reach thermal equilibrium. That is to say, the temperature controller reading may not be the actual temperature of the blackbody source during a transition period. This is a result of the fact that thermal sources are designed for maximum steady state accuracy. There will always be a time lag between actual blackbody temperature and controller reading. Although the lag exists (Figure 5), the average value calculated from the data is approximately the same. The command to change the source temperature should be "matched" to a temperature interval appropriate for the spatial frequency tested (i.e. the expected MRT), the slew rate and settling rate of the temperature controller. If the temperature interval is too small, the experiment will take too long and the observer will lose interest. If the temperature interval is too large, the target-background temperature difference will change too rapidly and the observer will not be able to effectively track his threshold.

Conclusions

The semi-automated method allows rapid assessment of the MRT value. The strip chart recording gives an immediate visual record of observer variability and can be used to train observers or, if necessary, to eliminate poor observers. The method is more accurate than the classical method because it forces the observer to track his own threshold. If the observer tracks his threshold with 30 up and down excursions then he defined his threshold 30 times. Whereas in the classical method he may define his threshold only once or twice at each spatial frequency. Furthermore, threshold is usually defined as the 50% probability of detection. A single event (classical method) is insufficient to accurately define a 50% probability point.

For automatic test equipment it is desirable to have a fully automated MRT technique to eliminate inter- and intra-observer variability. To eliminate variability, a criterion must be chosen which relates subjective detection to physically measureable quantities. The relationships among noise, signal and the eye-brain interpretation of four bar detection are not well understood. It appears that an experiment to elicit a criterion may be system specific and not universal.

The semi-automated method in part satisfies the automated requirement in that it immediately identifies observer variability. With any test, the data analyst must decide what variability is allowable. The threshold tracking represents a challenge to the observer and therefore he usually becomes highly motivated to minimize his variability, reduce his learning curve time and reach threshold in the minimum amount of time. Thus the MRT task which often is considered drudgery with the classical method becomes a challange with the semi-automated method.

Acknowledgement

This reseach was supported by Aeronautical System Division, Wright Patterson AirForce Base, Dayton, Ohio under USAF contract #F33657-80-C-0441.

References

1. Introduction to Statistical Analysis, W.J. Dixon and F.J. Massey, McGraw Hill, New York 1957, Chapter 19.

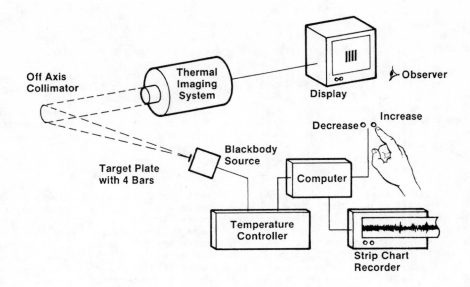

FIGURE 1.   METHODOLOGY FOR SEMIAUTOMATIC MRT TEST

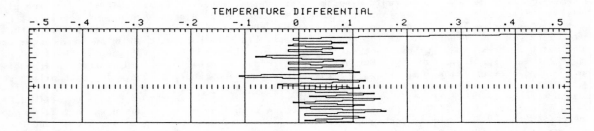

2A.   HOT TARGET

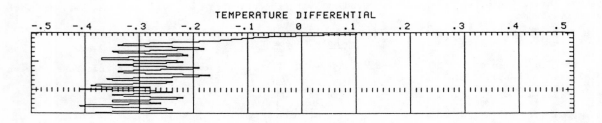

2B.   COLD TARGET

FIGURE 2.   TYPICAL DETECTION TRACES.   MINOR TICK

MARKS REPRESENT 10 SECONDS

TEMPERATURE DIFFERENTIAL

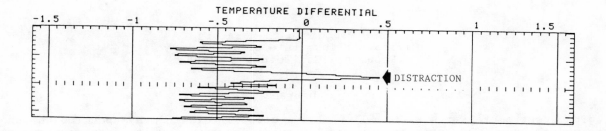

FIGURE 3. TYPICAL CURVE OBTAINED WHEN THE OBSERVER IS DISTRACTED

TEMPERATURE DIFFERENTIAL

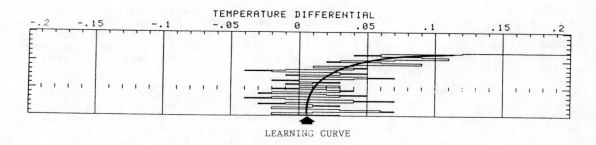

FIGURE 4. TYPICAL "LEARNING" CURVE

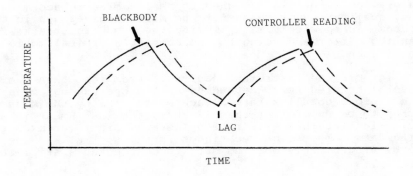

FIGURE 5. REPRESENTATIVE TEMPERATURE CONTROLLER

OUTPUT VS. BLACK BODY TEMPERATURE

An imaging spectrometer for the investigation of Mars

James E. Duval

Jet Propulsion Laboratory
California Institute of Technology
Pasadena, California   91109

Abstract

Imaging spectrometers are expected to play a crucial role in many upcoming unmanned planetary missions. The scientific goal of imaging spectrometry is to obtain compositional data about the target body by measuring the intensity and distribution of characteristic spectral signature features in the visible and infrared radiation reflected and emitted by surface materials. As part of the Planetary Instrument Definition and Development Program (PIDDP) sponsored by the National Aeronautics and Space Administration (NASA), a modular imaging spectrometer design with applicability to a diverse range of future planetery missions is being developed at the Jet Propulsion Laboratory (JPL). Derived from the near-infrared mapping spectrometer (NIMS) developed by JPL for the Galileo mission to Jupiter, the first version of the visual and infrared mapping spectrometer (VIMS) has been tentatively selected for NASA's recently approved Mars Observer mission.

Introduction

In the early 1990s, NASA will launch the Mars Observer mission, the first in a series of low-cost planetary missions to targets in the inner solar system. JPL will manage the Mars Observer mission for NASA in addition to supplying several of the science instruments. Representing NASA's return to Mars after more than fifteen years, the Mars Observer mission will be an important follow-on to the highly successful Mariner and Viking missions of the 1960s and 1970s.

The imaging data returned by the Mariner and Viking orbiters represent an important source of information for our understanding of the nature and evolution of the Martian crust and atmosphere. The imagery show clear evidence for a wide variety of geological processes, although details on the chemical and mineralogical properties of the geologic units, the inventory of the volatiles, and the rates of the processes involved are unknown. The scientific goal of VIMS on the Mars Observer mission is to utilize imaging spectrometry to obtain this mineralogical, chemical, and physical information, permitting a major step in understanding the nature and evolution of Martian geology.

The baseline VIMS instrument for the Mars Observer mission (figure 1) has 320 spectral channels with about a 12.5-nm sampling interval between 0.3 and 4.3 µm. The instrument's instantaneous field of view (IFOV) is designed to be 1.8 mrad to accommodate the spatial resolution and signal quality requirements for the Mars Observer application. "Whisk broom" imaging is accomplished with VIMS by moving the IFOV crosstrack with the instrument's integral scanning mirror and using the spacecraft's orbital motion to provide the downtrack direction of scan.

The internal data rate of the instrument can range from less than one kilobit per second (kb/s) to several hundred kb/s, depending on the operational mode and mission scenario. Since the typical downlink rate allocated to VIMS by the Mars Observer spacecraft is expected to be no greater than 32 kb/s, the instrument is designed with the capability of reducing its internal data rate by several orders of magnitude. This is accomplished by using data compression techniques in the spatial and spectral dimensions, data buffering, and by programmably editing data to change the spatial and/or spectral resolution.

Science objectives

The central objective for VIMS on the Mars Observer mission will be to acquire a global map of the surface lithologic materials based on their spectral reflectance properties. These data will provide compositional and mineralogical information for surface materials from a variety of absorption features that exist in the visible and near-infrared part of the spectrum. Most laboratory work has been done in the visible and near-infrared out to about 3.0 µm, so most known features occur there. These features occur in a variety of pyroxenes, clays and alteration products, hydroxyl groups such as in amphiboles, and chlorites. Distributional maps of these materials will be important for the interpretation of measurements made by other instruments. Additionally, the visible and near-ultraviolet part of the spectrum (extending down to about 0.3 μm) is important for Mars, as ferrous and ferric minerals, probably occurring as oxide rings on major lithologic units, are distinguishable in this spectral region, particularly in the dark Martian equatorial belt.

An additional scientific objective for VIMS is to study the formation, distribution, and migration of surface and atmospheric volatiles, particulary in the polar regions. As a result of the Mars Observer spacecraft's polar orbit, swaths on several adjacent orbits will overlap in the polar regions, allowing the composition and physical character of the polar caps to be monitored in detail during their seasonal advance and retreat. The principal information will come from near-infrared absorption features between 2.8 and 3.3 μm in which carbon dioxide and water, as found in ices and frosts or absorbed on surface materials, can be distinguished.

Instrument description

Table 1 summarizes the general VIMS instrument characteristics. The technique by which VIMS acquires multispectral imagery is illustrated in figure 2. "Whisk broom" imaging is accomplished by moving the instrument's IFOV crosstrack with an integral scanning mirror and using the spacecraft's orbital motion to provide the downtrack direction of scan. The pixel size is 650 m for an altitude of 361 km. The crosstrack scan will produce a swath containing 64 pixels for 41.6 km total swath width. The energy collected by the scanning foreoptics is spectrally dispersed in the spectrometer and sensed by 320 photodiode infrared and visible detectors. The instrument electronics processes and formats the detector signals for transfer to the spacecraft data system, where the data are recorded and/or transmitted to Earth. Ground data processing is ultimately used to produce spatially resolved spectra (multispectral imagery) from which surface composition information may be determined.

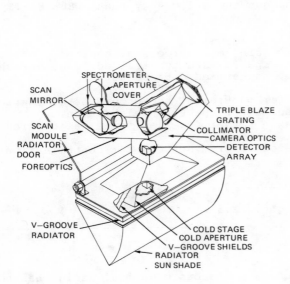

Figure 1. VIMS instrument

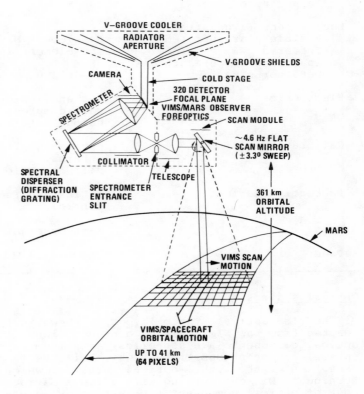

Figure 2. VIMS imaging technique

Table 1. VIMS instrument characteristics

Characteristic	Value
Heritage	Galileo near-infrared mapping spectrometer (NIMS)
Spectral range	0.3 to 4.3 μm
Spectral sampling interval	12.5 nm (average)
Pixel size	650 m from 361-km Martian orbit
Swath width	64 pixels for a 41.6-km swath from a 361-km orbit (6.6-degree FOV)
Scanner	linear, 4.6 Hz, full-aperture object-space (external) scanner operating at 80% scan efficiency
System AΩ	8.47 x 10^{-5} cm^2-steradian
Telescope	f/3.6, 21.0-cm focal length, 5.8-cm aperture, 1.8-mrad IFOV

Spectrometer:
  Entrance slit
  Collimator
  Disperser
  Camera
  Magnification
  Temperature
	
	374 μm (spectral direction)
	f/3.6, 42.0-cm focal length
	triple-blazed diffraction grating
	f/1.86, wide angle, flat field, 22.5-cm focal length
	0.536
	200 K

Focal plane (line array):
  Visible/near infrared
  Infrared
  Operating temperature
  Detector size
	
	64-element Si, quantum efficiency peak ≈50% @ 0.75 μm
	256-element InSb, quantum efficiency peak ≈80% @ 1.5 μm
	88 K
	200 x 200 μm

Electronics	3-channel signal chain, microprocessor control, software data compression, spatial and spectral editing, variable output data rate, packetized data output
Calibration	optical shutter for instrument background determination, internal line source for spectral calibration, diffuse reflectance target for flat-field calibration
Protective devices	optics and radiator covers, heaters for contamination control
Mass	Mars Observer = 18.71 kg
Power	day = 11.4 W; night = 2.5 W

Optical systems

The VIMS optical system layout is shown in figure 3. The basic optical design has been adapted from the NIMS/Galileo configuration. In general, the surface figures of the optical components are aspherics. Aluminum has been selected as the material for the optical component substrates, which will be fabricated by conventional machining with the exception of the actual optical surfaces, which will be produced by single-point diamond turning.[1] Diamond turning will allow a savings of approximately 25% to 30% over the cost of optical component fabrication using conventional optical shop methods. Utilization of aluminum as the material for all VIMS optical components will simplify the optical mounting interfaces in the structure. The monometallic design of the optics subsystem will allow for further simplification of the structure by eliminating the need for complex athermalization of the optomechanical subsystem.

Foreoptics. The VIMS foreoptics is an f/3.6 Ritchey-Chretien modified Cassegrain telescope. The telescope aperture is 5.8 cm and its focal length is 21.0 cm. Scanning the IFOV is accomplished by an external, full-aperture, flat mirror. In this configuration, the telescope performance is diffraction limited.

Spectrometer. The VIMS spectrometer is an all-reflective, Cassegrain system. The entrance slit of the spectrometer is coincident with the field stop of the foreoptics. An optical shutter is included at the entrance slit to the spectrometer to allow detector dark current and spectrometer background measurements. The dispersive element of the spectrometer is an aspheric, triple-blazed diffraction grating operating in the first order and located in the collimated light beam. The tripartide blaze arrangement is required due to the wide spectral coverage of the VIMS spectrometer, extending from 0.3 to 4.3 μm, or close to four octaves. Based on systems analysis of the instrument total throughput, 20% of the grating area is to be blazed at 0.45 μm, 20% at 1.4 μm, and 60% at 3.0 μm. The efficiency for the triple-blazed grating will be in the 70% to 80% range.

The dispersed spectrum is imaged on the detector array by a flat-field, all-reflective camera system that includes a calcium fluoride field-flattening lens. The magnification of the spectrometer optical system is selected in such a way as to form a spectral image of the entrance slit that exactly fills a 200 x 200-μm detector element.

Detector assembly

The VIMS detector assembly consists of a 64-element silicon (Si) photodiode line array for the visible and near-infrared (VNIR) band butted to a 256-element indium antimonide (InSb) photodiode line array for the short wavelength infrared (SWIR) band. Each detector element is 200 x 200 μm in size with a 30-μm dead space between detectors. The length of the entire array is 73.57 mm. The various components of the VIMS detector assembly are shown in figure 4.

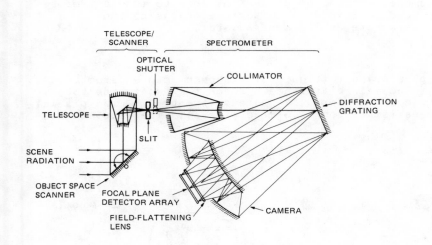

Figure 3.   Optical layout

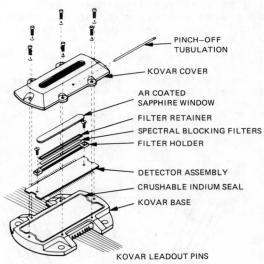

Figure 4.   Detector assembly

The Si photodiode array effective quantum efficiency is 50% peaked at 0.75 μm. Since the focal plane will be operated at the much lower temperature required for the SWIR array (88 K), negligible dark current can be expected for the VNIR array. The Si detectors are read out by a single 64-input multiplexer.

Prototype InSb line arrays have demonstrated excellent performance with high effective quantum efficiency (peaking at 80%), high uniformity and linearity (better than 3%), and low noise.[2,3,4] Cutoff wavelengths and dark currents typical for an InSb line array operated at 88 K are about 5.4 μm and 9.5 x 10^{-8} amp/cm^2, respectively. The InSb array is multiplexed in two groups of 128 detectors. The two multiplexers are laid out in an alternate "interdigitated" fashion so that the failure of a single multiplexer will not cause the loss of the whole spectral region.

An antireflective (AR) coating deposited directly on each array will allow both the VNIR and SWIR detectors to achieve higher effective quantum efficiencies. For the InSb array, each of the three pieces will have an AR coating tailored to achieve optimal quantum efficiencies in their respective operating spectral bands.

Spectral blocking filters are required to accomplish diffraction grating order sorting and to limit scattered light and background flux noise contributions to the signal. The transmission in the passband is greater than 70%, and is less than 0.1% in the stop band, virtually eliminating crosstalk from the higher diffraction grating orders.

Instrument cooling

The VIMS focal plane is cooled to 88 K using an advanced technology "V-groove" passive cooler (figure 5).[5] The cooler consists of a rectangular radiating aperture surrounded by a novel arrangement of lightweight, low-emittance, highly reflective radiation shields. Adjacent shields have an included angle of 1.5 degrees, creating V-groove cavities that face space. Thermal radiation incident on the cooler from the rear is intercepted by the shields and directed out the V-groove openings to space. The V-groove shields create a staging effect, but are much lighter than the massive metal plates used on conventional multistaged radiators. Feasibility testing of this cooler design has demonstrated the thermal isolation capability of the V-groove concept.[6]

The cooler radiating aperture is tilted parallel to the limiting planet ray and is shielded from the solar radiation by a sun shade. The sides of the radiating aperture are shaded from solar radiation and planet albedo by shields that are conductively isolated from the radiator aperture and V-grooves. The focal plane is attached directly to the radiating aperture cold stage structure, which is rigidly held in position relative to the spectrometer optics by a system of low-conductance filament-wound bands. The total radiating aperture area required for the VIMS cooler is 400 cm^2.

VIMS also requires that the spectrometer be cooled to 200 K or less to reduce thermal background radiation to an acceptable level. This will be achieved by using the space-facing area of the spectrometer structure as its own radiator.

Mechanical systems

Scan mechanism. The VIMS object space scanner (figure 6) has been adapted from the NIMS/Galileo internal scanner design. The main difference between the two is the total object space scan required, NIMS being designed for 0.57 degrees (twenty 0.5-mrad pixels) and VIMS being designed for 6.60 degrees (sixty-four 1.8-mrad pixels). The ±1.65-degree

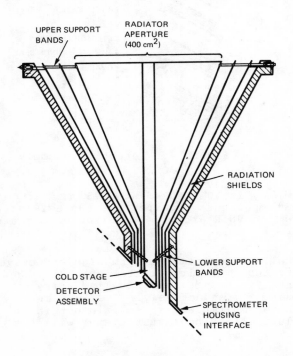

Figure 5. VIMS cooler

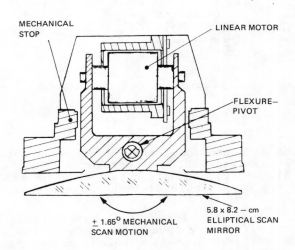

Figure 6. VIMS scan mechanism

motion of the scan mirror gives a total travel of +3.3 degrees. Since the foreoptics is in the same plane as the scan mirror motion, the +3.3-degree scan mirror travel is optically doubled to give a total effective scan of 6.6 degrees with no image rotation. Data are collected on the forward portion of the scan only, the backscan (or flyback) period being unused.

The scan mechanism is driven by a linear torque motor drive. The permanent magnet is attached to the scan mirror assembly, driven by the fixed solenoid, and pivots about flexures. The use of flexures provides long operational life without lubrication and also allows operation at near-cryogenic temperatures, if required. Position information for closed-loop control is provided by a linear voltage differential transducer (LVDT) attached directly to the scan mirror assembly. The scan mechanism has a demonstrated efficiency of greater than 80%.

Optical shutter. A simple, electromagnetically actuated, flexure-mounted optical shutter will be utilized by VIMS to block the entrance aperture of the spectrometer for instrument background and detector dark current measurements. It is required that the shutter cycle time be 20 ms or less so that at least four background integrations can be accomplished during a single scan mirror flyback period.

Optics and cooler covers. VIMS will be provided with reclosable covers to protect the optics and cooler from the effects of direct solar radiation and contamination. The actuator utilized for both covers consists of two independent direct-current motors coupled through a continuously engaged drive train to a common output shaft. Used on the Galileo spacecraft, these actuators are lightweight, compact, and free of common failure modes. The optics cover will be oversized and provided with a diffusely reflecting inner surface so that it may be opened 45 degrees to provide a flat-field calibration target for use when the spacecraft crosses the Martian terminator.

Structure. The use of aluminum optics allows for significant reduction in the complexity of the optics mounts and of the overall thermoelastic design. The structural foundation of the instrument is the spectrometer. A single-piece, machined-aluminum optical bench forms the backbone of the spectrometer. The foreoptics, the collimator, and the spectrometer camera are mounted to the optical bench. The diffraction grating is mounted via an aluminum sheet metal housing to the optical bench. The entire spectrometer is mounted to the spacecraft via a kinematic mount to minimize thermally induced distortion loading. The cooler (with the detector assembly) is accommodated by a set of trusses and attachment features on the optical bench and spectrometer housing.

Electronics

The electronics assembly orchestrates the operation of VIMS, directing data acquisition and processing, mechanism control, instrument monitoring functions, and communications with the spacecraft payload data system (PDS). A block diagram of the VIMS instrument and electronics is shown in figure 7.

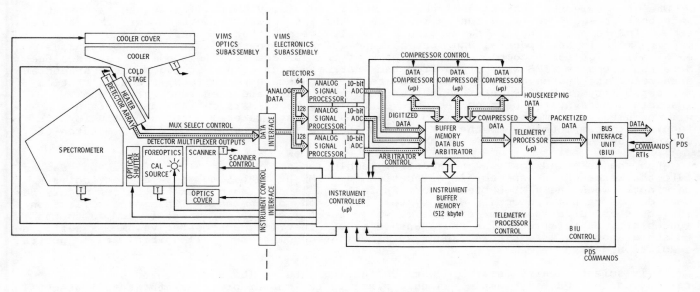

Figure 7. Functional block diagram

Analog signal processing. The signals from the 320 detectors on the focal plane assembly are multiplexed by three parallel analog signal processing chains. The 256 InSb detectors are read out by two interdigitated 128-to-1 multiplexers. The 64 Si detectors are read out by a single 64-to-1 multiplexer. The three analog signal processors, under direction of the instrument controller, apply individual gain and background corrections to each detector signal and then digitize to 10-bit accuracy. The most recently measured instrument flat-field and background calibration data for each detector element are stored in the instrument controller memory for this purpose. An automatic gain control feature is resident in the instrument controller software and is available to compensate for the large variation in solar illumination that will be encountered at different latitudes (a result of the spacecraft's sun-synchronous polar orbit).

Digital signal processing. As shown in figure 7, output from each of the three analog-to-digital converters (ADC) is transferred to the 512-kbyte instrument buffer memory. Once the required data volume (three complete spatial lines) is resident in the instrument buffer memory, one of the three parallel microprocessor-based data compressors accesses the data and executes a software version of the block adaptive rate controlled (BARC) data compression algorithm. It is estimated that three data compressors will be necessary to handle the maximum VIMS output data rate of 32 kb/s. Successfully utilized on both the Voyager and Galileo spacecraft, at least a 2-to-1 compression ratio is expected from the BARC process.[7] The compressed data are re-stored in the buffer memory where the microprocessor-controlled telemetry processor can next access them for reformatting into data packets. The transfer of VIMS data packets through the bus interface unit (BIU) to the spacecraft's PDS occurs under the control of the PDS on a real-time interrupt (RTI) basis. The maximum number of packets that may be transferred in any one second period is 8.

Resolution selection. Spatial and spectral resolution are selectable for VIMS to reduce the instrument's output data rate so that it can be accommodated by the spacecraft data system. Resolution reduction is simply accomplished by the deletion of undesired spectral and spatial information from the VIMS internal data stream, under the direction of the instrument controller, prior to gain and offset correction and digitation. In the spatial domain, this technique results in spatially undersampled imagery. Spatial sampling intervals of 1 (full spatial resolution) to 4 (every fourth crosstrack pixel in every fourth downtrack line sampled) may be specified, providing a symmetrical "checkerboard" sampling pattern. In the spectral domain, the number of spectral channels that may be selected is determined by the spatial resolution chosen and the desired output data rate.

Operational modes. The data generation rate for a particular VIMS operational configuration must fit the spacecraft data rate allocated to VIMS at that time in the mission. The spacecraft's maximum data transmission rate is expected to be no more than 32 kb/s (the "real-time" data link). Spatial and spectral resolution must therefore be selected such that VIMS will generate data at a rate compatible with the spacecraft allocation. The two basic modes of operation by which VIMS will collect virtually all of its spectral imagery data are the mapping mode and the snapshot mode.

When the selected spatial and spectral resolution for VIMS results in a data output rate that is equal to (or slightly less than) the spacecraft allocation for VIMS at that time, VIMS is operating in the mapping mode. It follows that the definition of the mapping mode is any spectral and spatial resolution combination that allows contiguous data collection for a complete dayside pass of the spacecraft. In the mapping mode, the data collected by VIMS are output from the instrument for either storage on the spacecraft data recording system or direct downlink to a Deep Space Network (DSN) tracking station. The capability of VIMS's internal buffer memory to store and then read out to the spacecraft data system data collected over the dayside of Mars during spacecraft night will be used to augment mapping mode operations. A data rate allocation of 1100 b/s is required to completely dump the VIMS data buffer during the approximately 59 minutes of darkness on the nightside of each orbit.

When the selected spatial and spectral resolution for VIMS results in an output data rate that exceeds the spacecraft allocation for VIMS at that time, data collection is terminated when the VIMS buffer memory is full and is continued when the buffer memory has been dumped to the spacecraft data system. This mode of operation, when data collection must be stopped for some period of time while the filled buffer memory is dumped to the spacecraft, is referred to as the snapshot mode. The VIMS data buffer volume is sufficient to store about 64 x 24 pixels (about 20 lines) of uncompressed full-spatial, full-spectral resolution data.

It is scientifically desirable when collecting data at low spectral and spatial resolution to regularly sample a single pixel in the center of the swath at full spectral resolution. VIMS has been designed with this capability to aid in identifying early in the mission regions in the spectrum that may merit closer study during future opportunities. This feature may be utilized in both the mapping and snapshot operational modes.

Instrument performance

The integration time for a pixel and the AΩ (area x omega) for the instrument determine the total energy that may be detected, subject to the limitation placed on the system by the total throughput. The instrument AΩ (the product of the area of the foreoptics aperture and the solid angle of the instrument IFOV) is 8.56×10^{-5} cm^2-steradians. The integration time (or dwell time) for a single pixel is determined by the scanner efficiency (assumed to be 80%), the number of crosstrack pixels (64), the pixel size on the Martian surface (650 m), and the orbital groundtrack velocity (3.0 km/s). For contiguous sampling downtrack, the scan frequency is required to be 4.6 Hz and the integration time is about 2.7 ms.

Throughput

The basic VIMS optical design is an all-reflecting Cassegrain system with a total of nine reflecting surfaces, including the diffraction grating. Three transmissive elements are included in the design, these being the field-flattening lens in the spectrometer camera assembly, the window to the hermetically sealed focal plane detector assembly, and the spectral blocking filters included in the focal plane detector assembly. The dominant central obscuration in the optical system is 40%, and is a result of the spectrometer camera secondary mirror. The diffraction grating is divided into three areas "blazed" for different peak response wavelengths in order to provide adequate signal over the broad spectral range VIMS is designed to cover. The quantum efficiencies of the two different detector materials utilized in the VIMS focal plane assembly (Si and InSb) complete the prescription for determining the total instrument throughput. Figure 8 depicts the "total throughput" of VIMS taking into account the above parameters. The relatively low total throughput is dominated by the reflectance off of nine reflecting aluminum optical surfaces (approximately 39%).

Signal contributions

The "total signal" and its major contributors (in electrons) are depicted in figure 9. The "solar source" curve is a result of sunlight (6000-K blackbody) at a distance of 1.52 astronomical units diffusely reflected from the Martian surface, which is assumed to have a spectrally uniform albedo of 30%. It is the "solar source" signal that contains surface material spectral signature information.

The "thermal source" is the thermal emission from Mars, modeled as a 240-K blackbody with 70% emissivity. Since the "thermal source" signal is not distinguishable from the "solar source" signal by VIMS, the sum of the two is plotted separately as the "source" and is considered to be the science data collected by VIMS.

The "spectrometer" contribution is a background signal that is a result of the detection by VIMS of its own thermal emissions. This internally generated thermal radiation is minimized by cooling the spectrometer assembly to 200 K, and the detection of any residual thermal radiation is lessened by optical baffling and the presence of the spectral blocking filters necessary to block unwanted grating orders. As a result of these measures, the only internally emitted thermal background occurs beyond 2.4 μm (due to the passband of longest wavelength blocking filter, ranging from 2.4 to 4.3 μm). The thermal emission of the foreoptics assembly is not significant because the energy must enter through the (small) spectrometer entrance slit and is then spectrally dispersed before it can be detected.

The "dark signal" (or detector dark current) is a background source and is the dominant contributing signal component for VIMS in the infrared. The "dark signal" for the Si detectors is negligible at the 88-K operating temperature of the focal plane array. For the InSb detectors, the "dark signal" is estimated at approximately 650,000 electrons (3 picoamps at 77 K, doubling with every 3-K rise in temperature). The "total signal" plot is the sum of all the above discussed signals:

$$\text{total signal} = \text{thermal source} + \text{solar source} + \text{spectrometer} + \text{dark signal} \qquad (1)$$

Signal-to-noise ratios

Every few scan lines during the scan mirror flyback time, the optical shutter is closed, and four detector array integrations and readouts are accomplished. The results of these four readouts are averaged and designated as the instrument background ("spectrometer" plus "dark signal"). The "source" signal (the science data) is determined by subtracting this measured background contribution from the observed "total signal" in the analog signal processor, prior to the ADC. The frequency with which the total background must be measured is dependent on the final detector operating temperature and the expected variation in focal plane temperature over the period of a Martian orbit. For a detector operating temperature of 88 K, an updated background measurement must be performed whenever a 0.01-K change in detector temperature is expected to have occurred. Current thermal analysis of the VIMS detector cooler assembly in the Mars Observer orbit indicates that this magnitude of temperature change may be expected to occur no more frequently than about every 12 scan lines on the dayside pass of the orbit.

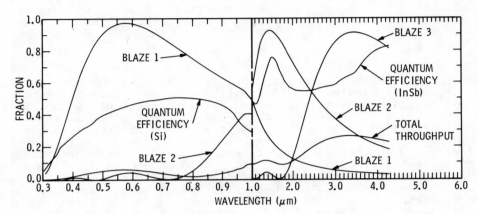

Figure 8. Total throughput

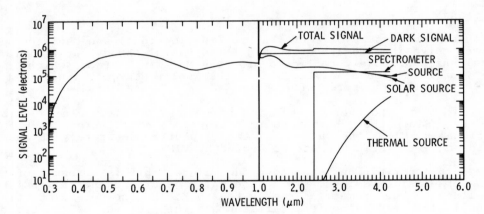

Figure 9. Signal levels

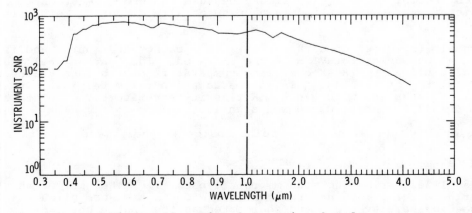

Figure 10. Signal-to-noise levels

Signal-to-noise predictions (figure 10) are approximated from the ratio of the determined "source" signal to the root-sum-square of the "total signal" shot noise and the measured instrument background shot noise:

$$SNR = source/\sqrt{(total\ signal) + (dark\ signal + spectrometer)/4} \qquad (2)$$

Acknowledgments

Most of the material presented in this paper has resulted from the work of others. The members of the VIMS Instrument Design Team and many others at JPL are acknowledged for having accomplished the work necessary to turn the VIMS instrument concept into an actual design. Additionally, I would like to thank Larry Soderblom, the VIMS Flight Investigation Team Leader for the Mars Observer mission, and the members of the VIMS Instrument Development Science Team for their help in the definition of the instrument concept. The development described in this paper is being carried out by the Jet Propulsion Laboratory, California Institute of Technology, under contract with the National Aeronautics and Space Administration.

References

1. Westort, K., "Diamond Machined Optics for Use at Visible Wavelengths," SPIE _Journal_, Vol. 508, 1984, page 2.

2. Bailey, G., Matthews, K., and Niblack, C., "Operation of Integrating Indium Antimonide Linear Arrays at 65K and Below," _Proceedings_ of SPIE, Volume 430, 1983, page 150.

3. Bailey, G., "An Integrating 128 Element InSb Array: Recent Results," _Proceedings_ of SPIE, Volume 345, 1982, page 185.

4. Bailey, G., "An Integrating 128 Element Linear Imager for the 1-5 μm Region," _Proceedings_ of SPIE, Volume 311, 1981, page 32.

5. Bard, S., "Advanced Passive Radiator for Spaceborne Cryogenic Cooling," _Journal of Spacecraft and Rockets_, Vol. 21, No. 2, March-April 1984, page 150.

6. Bard, S., "Test Report for the V-Groove Radiator Feasibility Experiment," JPL inter-office memorandum 3546-TSE-85-051 (JPL internal document), 22 April 1985.

7. Rice, R., "End-to-End Imaging Information Rate Advantages of Various Alternate Communication Systems," JPL Publication 82-61.

Invited Paper

Thermal Wave Imaging

Richard M Miller

Unilever Research Port Sunlight Laboratory, Quarry Road East,
Bebington, Wirral, L63 3JW

Abstract

Thermal wave imaging is a non-destructive testing technique which uses optically induced heat fluxes to probe the physical and chemical properties of solid samples. Localised heat sources are generated by the absorption of light from a focussed laser beam, and the propagation of heat from these sources is modified by the local thermal properties of the sample, providing a contrast mechanism. The technique can be applied to transparent, translucent and opaque materials. Information can be obtained about the thickness of coatings, the integrity of coating substrate interfaces, the presence of microscopic defects and inclusions, and the thermal characteristics of the material. Information can be obtained with a resolution of a few microns in both the X-Y plane, and along the Z-axis into the sample.

Introduction

Thermal wave imaging is a relatively recent addition to the armoury of non-destructive testing techniques which can be used for materials characterisation. It has developed out of the earlier technique of photoacoustic spectroscopy, and its particular characteristics offer potential for a wide range of practical applications.

In photoacoustic spectroscopy[1] a solid sample is contained within a gas tight chamber, one wall of which contains a sensitive microphone. The sample is illuminated through a window by an amplitude modulated light source, and an acoustic signal detected by the microphone. The signal is proportional to the amount of light absorbed by the sample, and by scanning the wavelength of the incident radiation, an absorption spectrum can be created. The generation of the photoacoustic signal can be understood from the scheme shown in Figure 1.

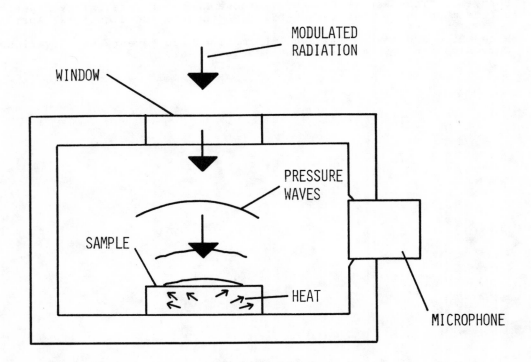

Figure 1.   The photoacoustic effect

If the sample absorbs some of the radiation falling upon it, excited states will be generated. In condensed phases these excited states normally have a very short life time, and providing the sample does not luminesce or undergo photochemical reaction, they will decay rapidly by internal conversion to generate a heat source modulated at the same frequency as the incident radiation, and localised within the sample in the region where the light was absorbed. Heat will propagate throughout the sample in a way controlled by the thermal characteristics of the material, and some of the heat will reach the sample surface. This modulated heat flux will be partially coupled by conduction into the surrounding gas, and will give rise to a localised expansion of the gas close to the sample surface. This 'acoustic piston' effect in turn produces a pressure wave which can be detected by the sensitive microphone. The photoacoustic signal can easily be recovered from background noise by a lock-in amplifier, as the only source of coherent signal at the modulation frequency of the light source derives from the thermal processes in the sample. As this is an energy conversion form of detection, absorption of light can be measured without having to ratio two beams of light as would be the case with conventional absorption or reflectance spectroscopy. As a result, the technique is able to study samples with specularly reflecting surfaces, materials of very high or very low optical density, and light scattering samples. Because of these properties, the technique has found a wide range of applications, principally dealing with samples which would be difficult to study by more conventional techniques.[1,2]

The photoacoustic signal depends on two properties of the sample; the optical properties of the sample which govern the amplitude and location of the modulated heat source, and the thermal properties of the sample which govern the propagation of heat.

Therefore, by replacing the wave length scanned light source with a focussed laser which can be spatially scanned across the surface of the sample, it would be possible to record this convolution of optical and thermal properties at each point in the scan, and build up a map or 'image' of the sample properties[3].

The key feature of the experiment, which differentiates it from many other imaging techniques, is that information is available not only in the X-Y plane of the sample, but along the Z-axis into the surface of the sample. This is due to the fact that an amplitude modulated heat flux behaves like a damped wave. This 'thermal wave' has a characteristic wavelength λ_t governed by the thermal diffusivity of the material and the modulation frequency of the incident radiation ω (equation 1).

$$\lambda_t = 2\pi \ (2\alpha/\omega)^{1/2} \qquad \text{Equ. (1)}$$

λ_t = thermal wavelength (cm), α = thermal diffusivity ($cm^2 sec^{-1}$),

ω = modulation frequency ($rad \ sec^{-1}$)

The wave is nearly critically damped, and does not propagate more than one to two wavelengths from the point of origin. By controlling the thermal wavelength through the modulation frequency, the zone of the sample which is probed can be varied.

Figure 2 shows how the various characteristics of the photoacoustic experiment can be used to provide imaging information. In Figure 2(a) a transparent coating with a step change in thickness is applied to an opaque substrate. An amplitude modulated laser focussed onto the sample will produce a localised heat source at the coating substrate interface. Heat will propagate through the coating to the surface of the sample where it will be detected as a modulated heat flux across the boundary. If the modulation frequency for the laser is sufficiently high the interface will lie more than two thermal wavelengths below the surface of the sample, the thermal waves will be damped before reaching the surface and no significant photoacoustic signal will be detected. If the thermal wavelength is long enough, the characteristics of the detected signal will be governed by the properties of the sample. Since heat propagates fairly slowly, there will be a phase shift in the signal introduced by the time it takes heat to propagate from the source to the surface. The signal amplitude will be a convolution of the optical properties of the substrate, governing the initial magnitude of the heat source, and the thermal characteristics of the coating governing the heat dissipation within it. In thermal wave imaging the signal is characterised by both the amplitude and phase.

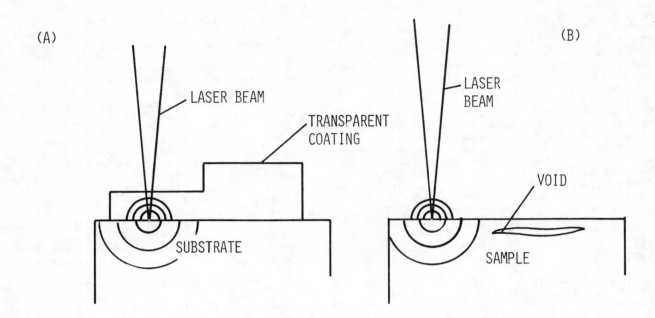

Figure 2.　The development of contrast in thermal wave imaging
　　　　　a)　transparent materials
　　　　　b)　opaque materials

As the laser scans across the step, there will be a change in the detected signal. The phase lag will increase due to the increased time taken for the heat to propagate through the thicker coating, and the amplitude will decrease due to the increase in volume of the coating involved. If one part of the coating is within the thermal diffusion length, and the other outside it, then more complex but equally dramatic changes will occur in both amplitude and phase. It is therefore clearly possible to obtain information about small changes in the thickness of a transparent coating.

Figure 2(b) shows an opaque material such as a metal, with a sub-surface void. When the laser beam is directed onto the solid metal, a localised heat source will be generated at the sample surface. Because of the thermal impedance mismatch between the sample surface and the surrounding gas, the heat will more readily propagate into the bulk of the material. This will give rise to a photoacoustic signal which is small in amplitude and displays only a small phase difference. As the laser moves over the void, the heat will be trapped in the thin region of metal above the void by the new impedance mismatch generated by the interface between the thin skin of metal and the sub-surface void. Heat will not be able to propagate readily into the bulk, and the signal amplitude will rise dramatically. In addition, since the heat must now escape by the relatively slow process of diffusing into the surrounding gas or into the void, a phase shift will be introduced into the signal.

If the modulation frequency were sufficiently high so that the thermal waves did not penetrate through the surface skin to the void, it would not be detected. Therefore from a knowledge of the thermal properties of a matrix, and the modulation frequency at which the void disappeared, it would be possible to estimate its depth within the sample. These two simple conceptual examples show that it is in principle possible to obtain depth relating imaging information about optical and physical structures in both transparent and opaque materials.

So far, only the gas microphone form of photoacoustic detection has been discussed. However, as it is only necessary to measure the time dependent variation in the sample surface temperature, a number of other detection schemes are possible. For example, a sensitive infra-red detector can be used to monitor the small fluctuations in emission on the surface as the temperature varies[4]. This technique has been successfully used for the quality control of thermal barrier coatings in industrial components[5]. In this case, only a single point measurement was required, but the approach is clearly applicable to imaging as well. Another method is to optically probe the thermally generated changes in refractive index in a fluid surrounding the sample[6]. Many other direct and indirect approaches are possible. A method which is significantly different in concept, and which has a number of practical advantages, is the use of a piezo-electric transducer clamped to the rear surface of a relatively thin sample[7]. The method is illustrated in Figure 3. As the optically induced thermal waves propagate, the induced temperature changes give rise to modulated stress waves within the material which propagate acoustically. The thermal waves generate acoustic waves whose amplitude and phase is related to the decaying thermal waves. Since the velocity of sound in most materials is very much greater than the rate at which heat diffuses, the acoustic waves have wavelengths many times greater than the thermal waves. They therefore propagate over relatively long distances without further interaction with the physical structures of the sample, and can easily be detected at the rear surface of the sample using a piezo-electric transducer. These are not optically induced acoustic waves, but a result of the optically induced thermal waves in the same way that the acoustic signal in the gas microphone cell is generated by the thermal processes in the sample. As a result, the acoustic waves detected by the piezo-electric transducer give information about the thermal wave processes, and the resolution of the imaging experiment is preserved.

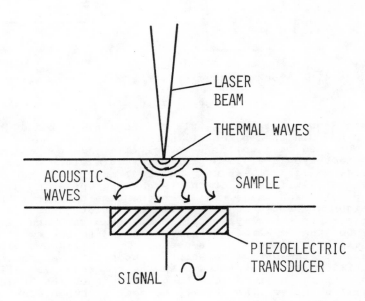

Figure 3.   Piezo-electric detection of thermal waves

Instrumentation

The instrumentation for thermal wave imaging is relatively simple. All that is required is an amplitude modulated focussed laser beam which can be scanned relative to the surface of the sample, and the detection system. Either the laser beam or the sample can be scanned, depending on user requirements. A block diagram for a typical experimental arrangement is shown in Figure 4. The laser is amplitude modulated by a mechanical chopper, or an acousto-optic modulator. The modulated beam is expanded and spatially filtered before scanning. The scan is accomplished by a simple arrangement of two orthogonal front surface mirrors mounted on limited rotation galvanometer motors. The deflected beam is then focussed onto the sample which may be in a gas microphone cell, or attached to a piezo-electric transducer. The signal from the detector is recovered using a two-phase lock-in amplifier. A computer controls the laser modulation frequency, the beam scanning, data acquisition and image generation. This instrument was designed for a maximum scan area of 1 cm^2 with a spot size of approximately 20 microns[8]. The examples shown in the following section were obtained using this instrument.

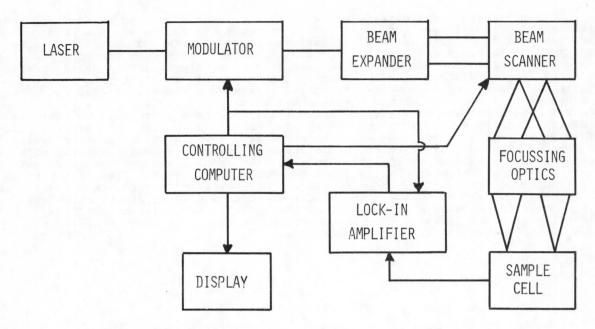

Figure 4.   Block diagram of a thermal wave imaging system

<u>Applications</u>

The image shown in Figure 5 is an example of the application of thermal wave imaging to the characterisation of thin transparent films[8]. The sample was a target of concentric coloured rings photographed onto colour reversal film. In each ring one dye-layer of the film is activated. Each dye-layer is approximately 5 microns thick and the overall thickness of the gelatine layer is about 20 microns, supported on an acetate substrate 130 microns thick. The overall diameter of the target was approximately 3 mm. The image shown is a 64 x 64 pixel phase image, in which the time taken for the heat to propagate from the point of light absorption to the surface is plotted as increasing image density. Each of the three layers in the sample are clearly resolved, and assuming a constant thermal diffusivity for the gelatine layer, correspond to the expected depths of maximum absorption for the different layers. The use of a phase image avoids the complication in the amplitude image where the optical absorption properties of the different chromophores at the laser line utilised are convoluted with the thickness variations. This sample demonstrates the ability of the technique to provide depth result information about complex multi-layer structures. The ability to monitor sub-micron fluctuations in film thickness has also been demonstrated[6,9].

Figure 5.   Phase image of multi-
layered thin film

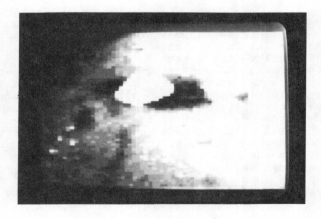

Figure 6.   Phase image of ceramic thermal
barrier coating showing failure

Figure 6 shows the phase image obtained from a steel test piece coated with a ceramic thermal barrier coating. This represents an example of a translucent sample. The coating is attached to the substrate with an intermediate bond coat to match the thermal expansion characteristics. Under repeated temperature cycling, these coatings may fracture and fail. The failure mechanism appears to be through oxidation of the bond coat layer which expands and cracks the overlying ceramic coating. Non-destructive techniques to study the on-set of failure are urgently needed. Figure 6 is a 64 x 64 pixel phase image of a test piece which has undergone catastrophic failure. The signal was detected using the piezo-electric transducer method. The large white patch in the image is the site of the failure, where a part of the coating has fallen away. Immediately either side of this zone are two dark regions which correspond to areas where the bond coat has undergone substantial oxidation, and has sites of incipient failure. Scattered around the image are other dark features which correlate with microscopic optical features on the coating surface, but which cover much larger areas. The technique is now being evaluated for use in the study and prediction of failure in this type of coating.

The final example is of a flame sprayed zinc coating on an aluminium substrate[10]. The average coating thickness was approximately 300 microns. Instead of a sub-surface delamination defect, a thickness defect was introduced by shadowing part of the sample during spraying. This produced a region in which the average coating thickness was approximately 140 microns. Figure 7 shows a 64 x 64 pixel phase image of the defect area obtained using a piezo-electric transducer. The thinner region of the coating is clearly seen as the lighter coloured area in the centre of the image. An interesting feature of this image was the very dark spot at the apex of the defect region. Although there is a dramatic increase in the phase lag, this is not due to a local increase in coating thickness. Optical examination of the sample shows that there is no region thick enough to produce this sort of phase lag. Clearly there is some sub-surface defect present, but it is not possible to determine what simply on the basis of the thermal wave image data. Figure 8 shows a close-up thermal wave phase image of the dark spot in Figure 7. Figure 9 shows a scanning electron micrograph of approximately the same region of the sample. A small crack in the coating can be clearly seen in Figure 9 at the apex of the defect. This crack appears to extend back under the main body of the coating. This is consistent with the image obtained in Figure 8. The dark region of the image matches the outline of the crack at its lower edge, but extends over a much greater area of the image. From these combined measurements, it is possible to determine the area covered by the void in the coating.

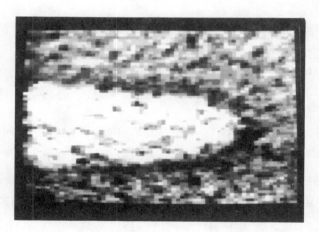

Figure 7.   Phase image of thickness defect in flame sprayed zinc coating

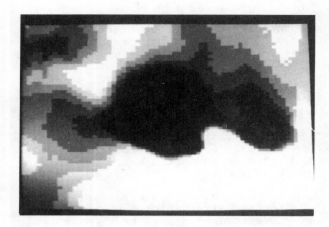

Figure 8.　High resolution phase image of localised defect

Figure 9.　Scanning electron microscope image of defect region shown in Figure 8

Conclusions

From the examples given above, it can be seen that thermal wave imaging has the potential to be a useful and versatile tool in the non-destructive evaluation of a wide variety of materials. Progress has been held up by the fact that most of the instruments have been constructed and used in research laboratories. Commercial instrumentation is now becoming available, and this will encourage wider practical application of the technique.

References

1.　A Rosencwaig, "Photoacoustics and Photoacoustic Spectroscopy", Wiley-Interscience, New York, 1980.

2.　J F McClelland, Anal. Chem., 1983, 55, 89A.

3.　A Rosencwaig, Science, 1982, 218, 223.

4.　P E Nordal and S O Kanstad, Infra-red Phys., 1985, 25, 295.

5.　H Reiter, A Almond and P Patel, Proc. 2nd Natl. Conf. Thermal Spray, Ohio USA, 1985, 119.

6.　W C Mundy, R S Hughes and C K Carniglia, Appl. Phys. Lett., 1983, 43, 985.

7.　G Busse and A Rosencwaig, Appl. Phys. Lett., 1980, 36, 815.

8.　G F Kirkbright, M Liezers, R M Miller and Y Sugitani, Analyst (London), 1984, 109, 465.

9.　G F Kirkbright and R M Miller, Analyst (London), 1982, 107, 798.

10. G F Kirkbright, M Liezers and R M Miller, Spectrochimica Acta (B), (In Press).

Invited Paper

Full field stress analysis using the thermoelastic principle

Lionel R Baker

Sira Ltd
South Hill, Chislehurst, Kent BR7 5EH, UK

and

David E Oliver

Ometron Inc
380 Herndon Parkway, Suite 300
Herndon, Virginia 22070, USA

Abstract

This paper describes a new method of measuring stress on the surface of a structure subjected to dynamic loading. Although the theoretical relationship between the changes in temperature at a point on the surface and the related change in stress has been known for over 100 years, it has only recently been possible to measure reliably and without contact the small changes in temperature down to 0.001°K involved. The performance characteristics of an instrument for Stress Pattern Analysis by measurement of Thermal Emission (SPATE) and the results of measurements on real structures recently obtained will be described.

Introduction

The stress distribution on structures in service can critically affect their life and integrity during use. Validation of new designs is therefore needed at an early stage before production resources are committed. Instrumentation, which has recently become available to map dynamic stresses in structures, allows the engineer to validate designs experimentally and with little effort.

The thermoelastic effect, which is a property of all materials, now provides the stress engineer with a facility for producing full field stress maps on the surface of structures. Within the elastic range a material subjected to tensile or compressive stresses experiences a reversible conversion between mechanical and thermal forms of energy. Provided that adiabatic conditions are maintained, the relationship between the reversible temperature change and the corresponding change in the sum of the principal stresses is linear and independent of the loading frequency.

The availability of high sensitivity IR detectors coupled with fast electronic signal processing systems has led to the development of a new stress measuring tool, based upon the thermoelastic effect, which has been named SPATE (Stress Pattern Analysis by measurement of Thermal Emission). This system detects the infrared radiation emitted from the surface of structures as a result of the small cyclic temperature changes which occur in a periodically stressed structure. The equipment can achieve a temperature discrimination at room temperature of 0.001°K which is equivalent to cyclic stresses of 0.4 N/mm^2 (58 psi) in aluminium and 1 N/mm^2 (145 psi) in steel with a spatial resolution down to 0.5 mm^2. The system, which provides full field stress maps containing several thousand individual stress points in a few minutes, will be described in what follows and its performance will be illustrated by a number of recently studied industrial applications.

Theory

The thermoelastic effect on solids is very similar to the cooling or heating of suddenly expanded or compressed gases. Within the elastic range theory [1] indicates the following relationship between change in stress $\Delta\sigma$Pa and the resulting change in temperature ΔT K.

$$\Delta T = \frac{-\alpha T \Delta\sigma}{\rho C\sigma} = -K_m T\Delta\sigma \qquad (1)$$

where T K   is the mean temperature at the measuring point

α K^{-1} is the linear thermal coefficient of expansion of the test material

ρ kgm^{-3} is the density of the test material

$C\sigma$ J $kg^{-1}K^{-1}$ Specific heat of the test material at constant stress

Km Pa^{-1} Thermoelastic constant of the test material $(= \alpha/\rho C\sigma)$

Most ordinary materials have positive values of α and K_m and so they are cooled by tension and heated by compression. A few average values for K_m are listed in Figure 1.

Material	$K_m(m^2 N^{-1})$	Measurable stress change with SPATE $(N\ mm^{-2})$
Steel	3.5×10^{-12}	1.1
Aluminium	8.8×10^{-12}	0.4
Titanium	3.5×10^{-12}	1.1
Epoxy	6.2×10^{-11}	0.054

Figure 1. K_m values for 4 materials and stress sensitivities.

Although the change in temperature could be measured by means of a contacting thermocouple it is clearly desirable to measure the small changes in surface temperature by means of a high sensitivity non-contacting radiation pyrometer. Figure 2 indicates schematically the cyclic loading of a plate with a hole and the associated time varying stress and temperature changes existing at the position of the black dot.

The total radiant energy flux, ϕ, emitted from a surface is related to the surface emittance, e, and the fourth power of the absolute temperature of the surface, T, by the Stefan-Boltzmann law [2] which states that

$$\phi = eBT^4 \tag{2}$$

where B is the Stefan-Boltzmann constant. It follows by differentiation that the flux change, $\Delta\phi$, resulting from a small change in the surface temperature, ΔT, is given by

$$\Delta\phi = 4eBT^3 \Delta T \tag{3}$$

If this flux change is recorded by a linear detecting system, the signal S will be proportional to the change in temperature and therefore to the change in the sum of the principal stresses i.e.

$$S = 4ReBKT^4 \tag{4}$$

where R is the detector response factor, or

$$\sigma = AS \tag{5}$$

where the constant A is a "calibration factor" relating the change in the principal stress sum to the detector system.

Measurement of cyclic temperature change

In accordance with the schematic representation of Figure 3, SPATE detects the infrared flux emitted from points on an observable surface as a result of the minute cyclic temperature changes in a cyclically stressed structure or component. It works over a cyclic stress frequency range of 0.5 Hz to 20 000 Hz. This enables assessment of most structures under realistic service conditions.

SPATE can measure instantly at a single point on the structure under test, produce a linescan of typically 120 data points in 2 minutes or less, or produce full field stress maps in a few minutes. Additional single point and lines of data can be recalled from scans stored in the system's non-volatile memory.

A spot of white light projected onto the structure under test enables easy identification of the measurement point.

The size of the measurement spot, or effective gauge size, increases with distance from the structure under test. With SPATE 0.3 m from the structure the measurement spot is only 0.5 mm diameter, allowing fine stress detail to be measured. With 1.0 m between the scan unit and the structure the measurement spot is 1.2 mm diameter.

The scan unit contains a high performance infrared detection system, cooled by liquid nitrogen. It is the performance of this detector which gives the excellent sensitivity to cyclic temperature changes of 0.001°K. The infrared detection system works over a range of test structure surface temperatures. Surface temperatures of over 600°C have been successfully stress mapped.

Two computer driven mirrors are located in the scan unit to produce a raster type of scan. The scan line, or area, can be operator selected anywhere within a 25° field-of-view. The limits of the area or line to be measured are fixed using a handset attached to the scan unit, and the projected white light spot.

The number of measurement points, data acquisition time and display mode are all entered through a fixed keyboard following a "user friendly" set of computer prompts.

The detected infrared signals are correlated in frequency, magnitude and phase with a load derived reference signal as indicated in Figure 4.

The frequency correlation ensures that measurements are made at the load signal frequency only, thereby eliminating environmental effects. The magnitude correlation gives an output proportional to $\Delta\sigma$, and the phase correlation determines the sign of $\Delta\sigma$.

This method of operation also provides a real time stress value which allows the stress engineer to measure a large number of discrete identified points very quickly.

Accuracy of measurement

Results from tests carried out using SPATE have been compared with theoretical and experimental data obtained by other techniques.

A well documented sample consisting of a mild steel plate with a circular hole in the centre, was sinusoidally loaded between 0.5 kN and 10.5 kN at 15 Hz. The specimen was scanned and a stress map produced. This stress map showed the region of high tensile stresses at the edge of the hole. It also showed a region of relative compression at the base of the hole, thereby illustrating the fact that SPATE can provide modal information.

Comparative work was undertaken using finite element methods and the analytical infinite plate solution. There is very close agreement between all three methods as shown in Figure 5.

When a correction is applied to the analytical infinite plate solution to allow for finite plate width and finite hole diameter, then it can be seen that there is almost perfect agreement with SPATE data.

Applications

The range of applications of SPATE extends to virtually all engineering materials and from small components, such as turbine blades, through to massive civil engineering structures. The work described so far has been in relation to a simple test specimen being loaded at uniform frequency and amplitude. SPATE can also be used on complex structures and under complex loading conditions.

For complex shapes, the non-contacting nature of the instrument means that, provided a line-of-sight can be established between the test structure and SPATE, stress data can be obtained. There is no need to view a structure normal to the surface, as in photoelastic techniques. Experimental work has established that there is no significant attenuation of the signal at 70° from normal to the surface. Similarly because of the long wavelength infrared radiation, in the range 8 - 14 µm, a much greater depth-of-focus is obtained than at visible wavelengths. These factors together means that complex shapes, curved surfaces and assemblies can be analysed, together with manufacturing features such as fusion weld fasteners, holes and changes in section.

Complex loading can be accommodated directly by SPATE. Running engines for example can be used to generate a cyclic load. A narrow band pass filter is then used to select specific frequencies for the load-derived reference signal and to determine the stress distributions associated with those frequencies of dynamic load. This technique has been found to be particularly useful for determining "dangerous frequencies" for structures. A typical example of work using a running engine to generate dynamic loading is shown in Figures 6 and 7.

Here the engine of a fork lift truck was run at various speeds and specific frequencies selected to determine the stress distribution around a headlamp bracket. In this real situation in which a manufacturer had experienced service failure, SPATE revealed the highly localised nature of the stress distributions (so localised that it had been missed using strain gauges) and was able to quantify the magnitude of these stresses.

It is interesting to note that because SPATE is both non-contacting and rapid, it is possible to effect practical modifications to a component such as this bracket and to quickly determine the effectiveness of the remedial action.

Comparison with other experimental methods

Strain Gauges

This method relies on the change of resistance of the gauge brought about by the deformation of the material to which it is attached. The technique requires gauges to be securely bonded to the structure under test. Data can then be obtained from each of the several gauges applied to the structure which can be loaded statically or dynamically.

Strain gauging is probably the most widely used experimental technique and operating conditions, corrections and calibration factors are well documented. However, the method does require lengthy surface preparation for gauge application. Also the positioning of gauges can be critical since highly localised stress concentrations may not coincide with the location of the gauge.

Strain gauges have a good sensitivity (around 5 $\mu\epsilon$) and can produce an adequate level of repeatability, provided that care and good experimental practice is adopted for gauge attachment. Also the capital cost of strain gauges is relatively low (although by the time a data acquisition system is included to scan several gauges the total costs can approach full-field systems). In relation to full-field methods strain gauges have poor spatial resolution, are not suitable for complex shapes, may modify the response of structures under test and may themselves be subject to mechanical failure.

Photoelasticity

Transmission model technique:

This technique requires the manufacture of a scale model in transparent plastic through which a beam of plane polarised light is passed. Under a static load an interference fringe pattern (isochromatics) will be visible in the model when viewed in a polariscope. The fringes must be counted from a datum to calculate the principal stress difference (proportional to the maximum shear stress) at a point in the model. By means of rotating the polariscope, taking further measurements (isoclinic evaluation), and going through a very laborious procedure the magnitude and direction of each principal stress can be determined over the entire surface.

A stress freezing technique, in which the loaded model must be heated (135°C) until the secondary bonds break down, can be used to retain the stress pattern within the model. Cooling the model and then sectioning it will allow a three-dimensional stress analysis to be undertaken.

The fringe patterns need skilled interpretation since it is not immediately apparent where the zero datum is located, and whether stresses are tensile or compressive. The model must, of course, be a very accurate representation of the actual component, and thus great care must be taken in its construction. Implicit in the modelling technique which, of necessity has to make use of a plastic rather than the actual structural material, is the fact that many practical situations such as anisotropy, inhomogeneity or components with individually moving parts cannot be modelled. Moreover, a precise knowledge of the loading conditions is necessary, and this is not always easy to evaluate or simulate.

Reflection photoelastic coating technique:

This technique is useful for surface stress measurement since there is no necessity to manufacture a scale model. A birefringent (e.g. double refracting) plastic coating is bonded to the structure with a reflective adhesive. The polarised light must, therefore, pass through the coating twice. The procedure for producing a uniform coating is rather laborious, time-consuming and messy. Moreover, only about one square foot (approx 300 mm x

300 mm) at a time can be covered. Results can only be obtained in areas accessible to light, and care must be taken to ensure that the coating does not reinforce the component. This technique is limited in field applications by the high level of skill and judgment required from the technician applying the coating. It is subject to errors due to ambient temperature and humidity fluctuations and other environmental effects, and suffers from the fact that the very bonding of a coating may stiffen or modify the structure and hence seriously affect the validity of the results obtained.

The maximum sensitivity which can be expected from photoelasticity is $\pm 10 \mu\varepsilon$.

Moiré interferometry

This is essentially a laboratory rather than a field technique more applicable to obtaining initial design information from model analysis rather than to the testing of engineering structures. If a diffraction grating is bonded to the surface of a component, a laser-interferometer containing a reference grid will reveal fringe patterns which are contours of constant displacement in a particular direction when the component is subjected to a load. The technique can only be used on a flat surface, and has an accuracy of ± 50 $\mu\varepsilon$ when a very high density grating (i.e. 10^6 holes/mm^2) is used; the accuracy is related to the hole density of the grating. The rather poor sensitivity of the Moiré technique means that it is more suited to plasticity investigations rather than the relatively low strain range normally encountered in experimental stress analysis. Moreover, the fact that the fringes are in the form of displacement contours rather than stress is a great disadvantage for stress analysis since there is no automated procedure available as yet for the conversion of the Moiré fringe patterns to stress. The Moiré technique has, however, been used to track damage growth in composite materials, to evaluate fracture mechanics parameters, and to indicate the non-uniform nature of residual stress fields. The technique is only applicable to static loading.

Laser Holography

This technique consists of recording photographically the diffraction pattern produced by laser illumination of a component, and later reconstructing its image from the diffraction pattern. The image has depth but is not a complete 3-D description since it only reveals the parts of the component which were illuminated by the laser. The basic diffraction pattern is produced by recording the interference pattern between the laser beam and the light back-scattered from the component. A pulsed laser can be used for dynamic applications. The hologram stores information as to the intensity of the light back-scattered from the component and its phase relative to the laser beam. The phase information is related to the position of the component, and it is this which provides the three dimensional property of the reconstructed holographic image.

Like the Moiré technique the holographic pattern is in the form of displacement rather than stress contours, but it is insensitive to displacements which are other than out-of-plane. The technique requires complex and expensive equipment which must be set-up exactly in order to give accurate data. Under ideal conditions the theoretical accuracy will be better than 1 μm (i.e. similar to the wavelength of the laser). However, practical constraints mean that this high accuracy is unlikely to be achieved.

Brittle lacquer

This technique involves the spraying of a suitable brittle coating on to the surface of a component. The brittle coating will crack perpendicular to contours of maximum principal stress if the maximum principal strain exceeds the threshold value of the lacquer. The threshold strain is about 500 $\mu\varepsilon$ but depends on temperature and humidity, thereby making reproducibility difficult to achieve. It is a cheap and useful on-site technique if maximum tensile stresses are of primary interest, but it is imprecise, labour intensive, and is therefore normally used as a preliminary to other more precise techniques. Lacquers can be somewhat toxic and are largely insensitive to compressive stresses.

Thermoelasticity

Stress Pattern Analysis by measurement of Thermal Emission is a new technological development which is based on a hitherto unexploited physical phenomenon known as the thermoelastic effect. The thermoelastic effect refers to the thermodynamic relationship between changing stress in a component and the temperature changes in the component produced by those stress changes. Under adiabatic conditions, produced by dynamic loading, the change in temperature at a point in a component is proportional to the change in the use of the principal stresses at that point.

The advantages of the thermoelastic technique are:

* A full field stress pattern can be obtained very rapidly.

* Only minimal surface preparation is required (i.e. cleaning and spraying with high emissivity paint).

* Complex geometries can be evaluated almost as easily as flat surfaces since the thermoelastic response is not particulary sensitive to viewing angle, and the instrument has a large depth-of-focus.

* A spatial resolution of 0.5 mm can be achieved.

* The sensitivity of the instrument is higher than can usually be achieved using other full-field techniques. The stress sensitivity is dependent on the component material, but 0.4 MPa (60 lbf/in^2) can be attained in aluminium. Furthermore, the technique can differentiate between compressive and tensile stresses.

* The technique is unique in being a non-contacting stress measurement on a real structure.

* The field-of-view is altered simply by moving the camera and/or the limits of the scanning mirrors. Similarly the number of stress measurements within the field-of-view is preselectable by the operator.

* A colour coded stress pattern, displayed directly on a VDU, is simple to interpret.

* The stress patterns may be archived on a dedicated floppy disk unit or transferred to cartridge disk on a mainframe for detailed examination or post-processing.

* High temperature measurements of qualitative stress patterns may be undertaken.

* The system is capable in principle of extension to dynamic service loading applications.

* Results correlate well with both theoretical and other experimental methods and the system integrates readily to existing testing practice.

* In addition to full field data, individual point or line data can be produced with a consequent saving in data acquisition time.

* Experimental methods are available to accommodate displacements normal to the viewing direction (which would otherwise smooth the measurement stress data over the movement range).

* The technique works on virtually all structural materials including metals, composites, plastics and ceramics.

* Design or manufacturing features, such as welds, fasteners etc, can be assessed directly.

* No controlled environment is necessary for operation.

The disadvantages of SPATE are listed below:

o The system is relatively expensive - although the actual cost of producing stress data is small.

o For accurate quantitative work the loading must be dynamic at a frequency greater than 3 Hz for steels in order to achieve adiabatic conditions. The stresses must also remain within the elastic range of the material (isentropic conditions). However, good qualitative data can be obtained even when these conditions are not met.

o The stress pattern is in the form of principal stress sum. Individual principal stresses and directions cannot be separated. Pure shear produces no thermoelastic output.

o Only surface stress can be measured.

o Optical access is required (special infra-red mirrors can readily extend the field of access).

o Calibration from material properties requires a precise knowledge of physical data which may be difficult to obtain for the component material. However, a simple experimental 'known stress' method is available whereby a strain gauge rosette is used to provide calibration data.

Conclusions

The application of modern high sensitivity thermal radiation detectors in conjunction with specially designed electronic signal processing techniques has enabled the construction of a radiation pyrometer able to measure temperature changes down to 0.001°K. An instrument of this sensitivity, which employs the principle of thermoelasticity, is able to measure stress changes comparable with the smallest that can be achieved by a typical strain gauge on real structures and remotely. Extensive trials have confirmed the high accuracy of measurement possible with this technique and revealed the potential of the method to examine areas of structures not accessible by any other means. It may be assumed therefore that this new branch of infrared technology will have an increasing impact in the fields of mechanical and structural engineering.

References

1.   Thomson, W., (Lord Kelvin), Jnl Maths (1855), reprinted in Phil Mag, Vol. 5, pp. 4-27. 1878.
2.   Rogers, G. F. C. and Mayhew, Y. R., Engineering Thermodynamic Work and Heat Transfer, Longmann. 1967.

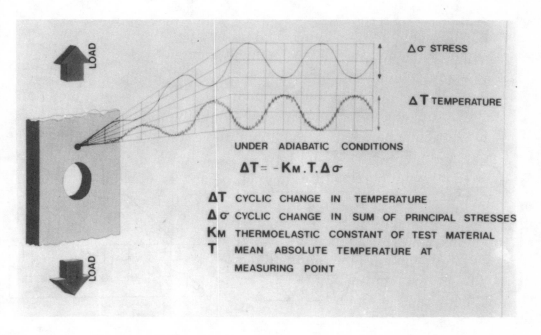

Figure 2. Thermoelastic theory for cyclically loaded structures.

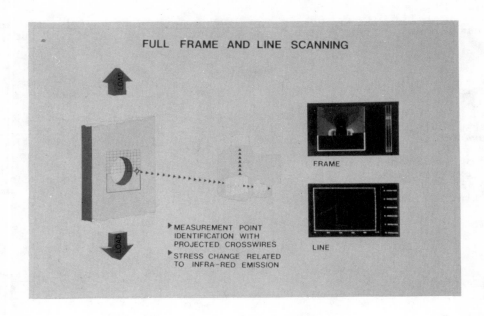

Figure 3.   Schematic representation of SPATE.

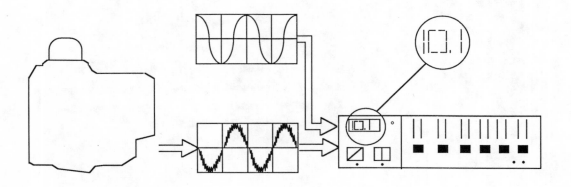

Figure 4.   Correlation between temperature and reference signals.

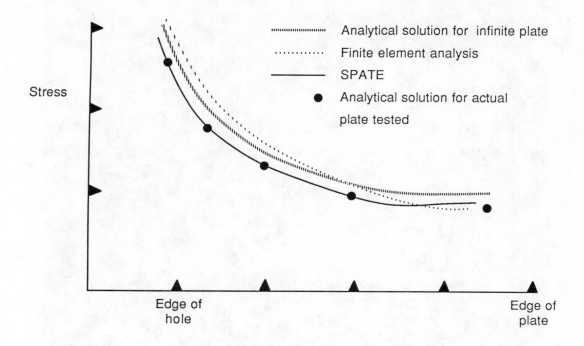

Figure 5.   Comparison of SPATE data with other methods.

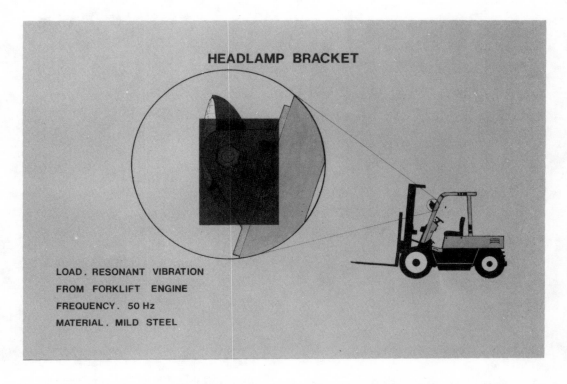

Figure 6.   Schematic showing area scanned on running fork lift truck.

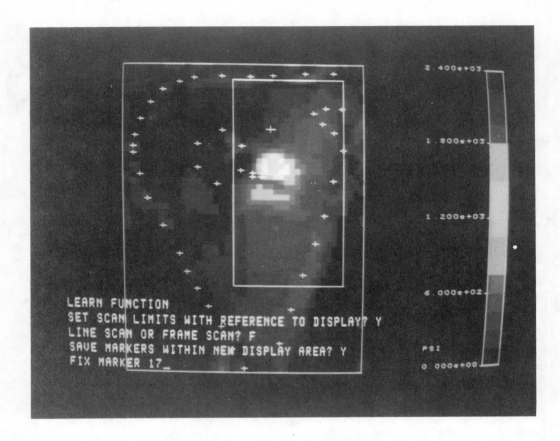

Figure 7.　Measured stress pattern from fork lift truck.

Invited Paper

Shuttle infrared imaging experiment

A. Aronson, R. Cenker, H. Gilmartin

RCA Astro-Electronics Division
P.O. Box 800
Princeton, N.J. 08543-0800

Abstract

The RCA Infrared Imaging Experiment (IRIE) flown January 12-18, 1986, on Shuttle Mission 61C is reported. The infrared camera, which was operated in the 3.5- to 5-μm spectral band, replaced one of the visible CCTV cameras on the Shuttle. The camera employed a 160 x 244 element monolithic platinum-silicide (PtSi) area focal-plane array developed at RCA Laboratories. The array characteristics, camera electronics, optics, and focal plane cooling are summarized.

The preplanned scenes for the IRIE are listed. A total of about 2.5 hours of data, including some preplanned scenes and unscheduled operation, were recorded on the first, third, fourth, and fifth days of the mission. Several of the recorded scenes are mentioned specifically in the paper, and a short videotape will be shown at the presentation. Quantitative analysis of the data is in progress.

Introduction

RCA has configured an infrared imager as a Shuttle experiment to test the performance of Schottky-barrier infrared focal plane arrays (SB FPAs). This Infrared Imaging Experiment (IRIE) was designed, built, and tested in eight months as an RCA-funded project. Central to the overall feasibility of the effort was an IRIE design that replaced an existing visible CCTV camera (also built by RCA) on Shuttle flight 61-C using the existing CCTV interface. Thus the IRIE was completely Shuttle-interface transparent.

Schottky barrier IR CCD technology, originally conceived in 1973 by Shepherd and Yang, is a candidate for spaceborne staring and scanning sensors requiring focal planes with 10^3 to 10^6 or more detectors.[1,2] SB FPAs are the lowest cost and most producible of current large-scale FPA technologies, and there is a good deal of interest in obtaining flight experience with these devices.

One of the authors, Robert Cenker, was the payload specialist for an RCA K-band communications satellite that was deployed during Shuttle flight 61-C. He also participated in the planning and operation of the IRIE flight experiment. The preplanned list of earth scenes for the IRIE included hot targets, such as volcanoes, fires, and urban areas; and geographical features, such as lakes, coast lines, and rivers. Celestial/orbiter scenes listed were Orbiter thermal surveys, earth's limb, airglow, and Satcom K1 deployment. These lists were approved by NASA and formed the basis for the operation plan for the IRIE.

The IRIE was very successful, and about 2.5 hours of imagery were recorded. Selected portions of these data will be shown during the presentation of this paper.

IRIE Camera System Description

The IRIE camera shown as it was mounted on the aft bulkhead in Figure 1, was based on an existing prototype developed at RCA Laboratories.[3] The IRIE was designed to fit the Shuttle pan/tilt CCTV camera mount and use the identical interface. A dimensioned drawing of the IRIE camera is shown in Figure 2. Table 1 summarizes the performance characteristics of the IRIE. The IRIE camera consists of an optical assembly, cooled Schottky barrier FPA, cooler, electronics, pan/tilt unit (PTU), and commands and telemetry; these are shown in the block diagram of Figure 3.

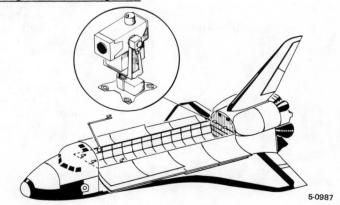

5-0987

Figure 1. IRIE camera in shuttle cargo bay.

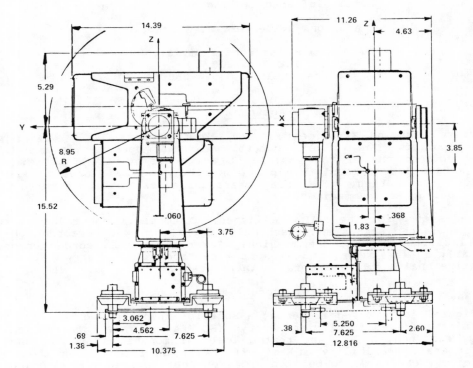

Figure 2. IRIE mounted on its PTU.

Table 1. IRIE Camera/PTU Characteristics

Parameter	Characteristic/Value
Lens	
• Focal Length	88 mm f/2.4, adjustable focus
• Ground FOV and IFOV	27.6 x 21.1 miles,* 1050 x 525 feet*
• Optical Pass Band	3.5 to 4.5 μm
Camera	
• Video Format	Compatible with EIA RS 170 525-line interlaced format. The IR CCD contains 244 vertical elements; alternate lines on each field are blank (black level).
• Sync and Commands	75-ohm balanced line input. Commands multiplexed with EIA sync during lines 11 and/or 13. 10 camera commands 4 lens commands 7 PTU commands
• Memory	8 x 40,960 bits; stores one frame as reference black.
• Video Output	75-ohm balanced line. Status and test signals present during five vertical lines.
• Input Power (nominal)	Camera Electronics: 38.1W Cooler: 25W (steady state)
• Input Voltage	28 V_{dc}, nominal
Pan/Tilt Unit	
• Travel Angle	±170° aximuth ±170° elevation
• Drive Type	90° stepper motor
• Step Angle	0.2° per step, pan or tilt
• Step Rates	1.2° per second or 12° per second
Weight	
• Camera	22.8 pounds
• PTU	13.0 pounds
• Cable	0.5 pound
*190-nmi orbit.	

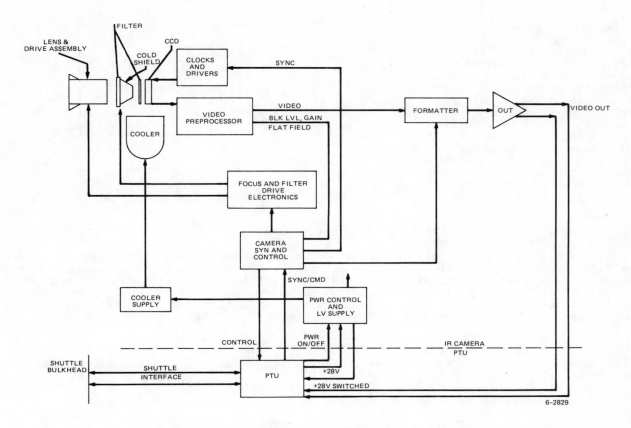

Figure 3. IRIE camera, block diagram.

Schottky Barrier Focal Plane Array

Schottky barrier detectors fabricated using standard silicon IC processing techniques are free from blooming and lag and exhibit extremely small (less than 5%) variation in responsivity across the array. Platinum silicide Schottky detectors offer useful sensitivity below 5 μm and are typically cooled to about 80K. They are also low noise and operate at very low power dissipation levels. These characteristics, together with high producibility, make this detector technology attractive for a wide range of staring and scanning space, ground tactical, and commercial infrared detection applications.

The responsivity and quantum efficiency with wavelength for a typical PtSi Schottky barrier detector is shown in Figure 4. Detectivity for a good SB detector is limited by dark current and background shot noise. D* versus background irradiance for typical SB detectors is shown in Figure 5.

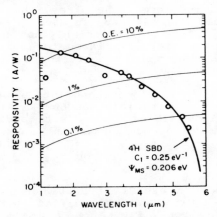

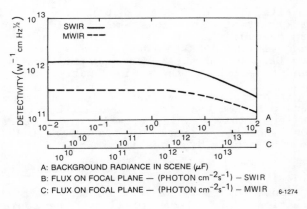

Figure 4. Responsivity of PtSi SBD fabricated with 20A of PtSi, 3-hour anneal time, and 6500A SiO$_2$ dielectric.

Figure 5. Detectivity (D*) vs. background irradiance for SB FPA.

The IRIE used a PtSi 160 x 244 pixel CCD FPA. The 160 x 244 pixel FPA is an interline transfer IR CCD designed to operate with a readout of two vertically interlaced fields per frame. An individual pixel format is shown in Figure 6. The PtSi active area of 50 μm x 25 μm shares real estate with the CCD readout electronics in this monolithic construction for an overall pixel dimension of 80 μm x 40 μm.

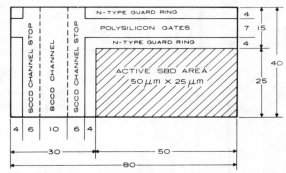

Figure 6. Simplified pixel layout of 160 x 244 IR CCD FPA.

Optical Performance

The optics of the IRIE consisted of an 88-mm f/2 variable-focus lens, infrared bandpass filters, and a cold shield that formed an f/2.4 cold stop. The calculated instantaneous field of view (IFOV) with an 88-mm focal length lens using a detector pitch of 80 x 40 μm was 0.9 x 0.45 mr; this provides a ground footprint of 1050 x 525 feet from an orbital altitude of 190 nmi. Measurements with a bar target at the Nyquist frequency showed a 64% peak-to-peak response compared to a low-frequency measurement.

Radiometric Performance

For extended sources, the noise-equivalent signal (NES) performance of the IRIE at 4 μm can be calculated using:

$$NES = \frac{4q(f/\#)^2 N_e}{R_\lambda t_{int} \tau_o \pi A \eta_{ff}} \qquad (1)$$

where q equals 1.6E-19, f/# is 2.4, R_λ is the responsivity of 0.02 A/W at 4 μm, t_{int} is the integration time of 33 ms, τ_o is the overall optics throughput of approximately 0.5, N_e equals 250 noise electrons with zero background, A is the detector area of 3.2 E-5 cm^2, and η_{ff} is a fill factor of 0.39.

Substituting these values in equation 1, NES = 7.15 E-8 $W/cm^2/sr$; this provides a noise-equivalent temperature difference (NETD) of 3.4K at a 200K scene. The IRIE was nearly background noise limited for an NETD of 0.08K at 300K scenes.

Electrical Design

The electrical design of the IRIE consists of typical CCD focal plane array TV techniques and produces standard RS 170 TV format output. These techniques are described in a paper to be published.[4]

Although the SB FPA exhibits a very uniform responsivity across the array, there are dark current variations among pixels, and lens and cold shield nonuniformities, which can produce fixed pattern noise or shading in the image. This is particularly noticeable when viewing 300K background scenes of people in a room or the earth where the contrast may be low and the video gain is usually high. A frame uniformity corrector is often used to eliminate these fixed patterns.

In operation, the frame corrector stores an 8-bit digitized video frame into a memory during a time when the IRIE is tilted to view a blackbody to irradiate the SB FPA with a uniform flux. During normal camera operation, this blank stored scene, with additive nonuniformities frozen in, is subtracted (pixel by pixel) from all subsequent frames to provide a very high quality image.

Cooler and Cold Shield

Closed-cycle mechanical coolers with useful lifetimes in excess of 1000 hours were practical for a 7-day mission. The cooler chosen was a Cryodynamics modified Stirling cooler. This cooler weighed 5.6 pounds with the vacuum shroud (Figure 7), had a 1-watt cooling capacity at 77K, and a minimum no-load temperature of 50K. Input power to the cooler averaged about 30 watts. Mechanical vibration from the cooler was insignificant, as the limiting resolution MTF did not change measurably when the cooler was switched on and off.

The cooler cold finger interfaced with the focal-plane array in a vacuum housing that contained the cold shield. The cold shield aperture formed an F/2.4 cold stop to shield the FPA from the lens barrel. A vacuum pump (not part of the IRIE) evacuated the vacuum housing for ground operation; a plug was manually pulled before launch to vent the enclosure to cargo bay ambient. Vapor traps were designed into the cold shield to prevent Shuttle-produced water vapor from freezing on the FPA.

The FPA temperature was maintained at 82K ±0.2K by regulating the cooler motor speed with a pulse width modulated (PWM) power source using a closed-loop temperature controller. The PWM reference was a temperature sensor bonded to the cold finger.

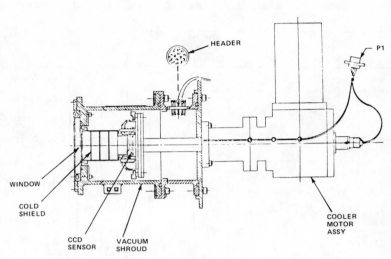

Figure 7. Cryogenic cooler assembly.

Infrared Imaging Experiment Mission

The IRIE was flown on Columbia Mission 61-C from January 12 to 18, 1986; and it was used to gather a wide variety of data, including orbiter cargo bay thermal scenes, geographic scenes, airglow, and thruster firings.

The primary responsibility of Payload Specialist, Robert Cenker, on Mission 61-C was to operate the IRIE. This camera has obvious applications for viewing earth resources and monitoring the payload bay. It was for these reasons that the camera was included on 61-C's mission manifest at a relatively late date in mission planning. Planning for the IRIE operations began with a list of tentative scenes drawn up by RCA Astro-Electronics. This list of scenes (Table 2) contained various views of the earth and space that provided a variety of thermal inputs, such as ocean currents, urban and rural areas with prominent geological features, and areas of volcanic activity. In addition to these clearly defined targets, several "scenes of opportunity" were identified. These included thunderstorms, forest fires, and any other similar events that might provide data as to the infrared camera's capabilities. Priorities were assigned on a scale of one to three, one being the highest. This list was subsequently entered into the PIP (Payload Interface Plan), a joint agreement between RCA Americom and NASA, which actually defined mission operations.

Many of the targets listed in Table 2 were observed. These included terminator data, sunglint, volcanoes, cities, orbiter RCS plumes, the Satcom deployment, and cargo bay imagery.

The IRIE operation provided a significant amount of data especially in Central America and Mexico. These areas were particularly well suited for observation due to the obvious nature of the landmarks in the area, making it particularly easy to identify and orient the camera to pick up items of interest. Data were accumulated over several days during the mission and included volcanic hot spots and geothermal activity.

The IRIE was particularly useful to the crew for thermal surveys of the cargo bay where it provided high resolution day/night imagery of cooling lines, transmitting antennas, "Get Away Special" modules, and other components.

RCA is continuing to study and evaluate the data to provide radiometric calibration and enhancement of the imagery.

Acknowledgments

Credit is due to the scientists and engineers from the RCA Laboratories, Princeton, NJ, Advanced Technology Laboratories, Moorestown, NJ, and the Astro Electronics Division, Princeton, NJ for their contribution to the development of the IRIE. The development of the 160 X 244 FPA was supported by RADC, Hanscom AFB, MA and many of the camera system concepts were jointly developed by RADC and RCA.

Table 2. Preplanned List of Scenes for Flight 61C

Scene	Quantity/Lighting Day	Night	Priority 1	2	3
1. North Africa desert/shoreline	1	1		X	X
2. Saudi Arabia desert/shoreline	1	1		X	X
3. Thunderstorms, hurricane	*	*		*	
4. Terminator - sunrise/sunset	2	2	X		
5. Sunglint-water and clouds	*			X	
6. Forest fires/volcanoes/gas flares	*	*		X	
7. Cities					
• San Juan	1	1	X		
• Suez Canal	1	1		X	
• Honolulu	1	1	X		
• Houston/Galveston	1	1		X	
• Rio di Janiero	1	1			X
• Miami	1	1	X		
• Singapore	1	2			X

Scene	Quantity/Lighting Day	Night	Priority 1	2	3
1. Moon				X	
2. Earth's limb		4	X		
3. Aurora		3*		X	
4. Zodiacal light		1*		X	
5. Orbiter RCS plume		3		X	
6. EVA/RMS end effector	*	*	X		
7. Satcom deployment	1 or	1		X	
8. Cargo bay	2	2**	X		
9. Surface glow		3	X		
10. Bay door closure	1 or	1			X
11. Other payload deployment	*	*			X

* Scenes of opportunity.
** Once with lights on, once with lights off.

References

1. F. D. Shepherd and A. C. Yang, "Silicon Schottky Retinas for Infrared Imaging," Joint Electron Devices Meeting, Tech Digest, pp. 310-313, 1973.
2. H. Elabd, A. Aronson, J. Tower, "Schottky Barrier IR-CCDs In Earth Sensing Applications" 1982 Government Microcircuit Applications Conference (GOMAC) Orlando, FL, 2-4 November 1982.
3. W. Kosonocky, F. Shallcross, T. Villani, and J. Groppe, "160 x 244 Element PtSi Schottky-Barrier IR-CCD Image Sensor," IEEE Transactions on Electron Devices, Volume ED-32, No. 8, August 1985.
4. J. Groppe, et al., "Imaging from the Shuttle Columbia using a Schottky Barrier Staring IR CCD Camera," Proceedings SPSE, 5th International Conference on Electronic Imaging, Oct. 14-17, 1986, Crystal City, Arlington VA.

Signal processing of infrared imaging system

Li layuan

Department of Electrical Engineering and Computer Science, Wuhan University of Water
Transportation Engineering, China

Abstract

The signal processing techniques of infrared imaging system are discussed. Performance
of PEV for chopping mode in the system and some basic designing principles of the system
are described. Main methods for processing signal of infrared imaging system are suggested.
Emphasis is laid on the multiple fields accumulation and image difference processing technique. On the basis of describing the main principle of the method, the concrete project
is put forward. Some test results are also given.

Performance analysis of PEV in chopping mode

With the influence of the modulation transfer function (MTF) neglected, the peak — peak
value of the PEV signal current in chopping mode can be expressed as

$$i_{sm} = \frac{PW}{2\rho cd} t_f V_B h , \qquad (1)$$

where P — pyroelectric coefficient,
W — power absorbed, per unit area of target,
ρ — density of pyroelectric material,
C — specific heat capacity of target material, d — thickness of the target, t_f — field period, V_B — scanning velocity, h — scanning line width of the beam. It is noted that W is
proportional to the power difference between the powers radiated respectively by object
and background per unit area, i.e.

$$W \propto \sigma [(T_o + \Delta T)^4 - T_o^4] \cong 4\sigma T_o^3 \Delta T .$$

where T_O is the background temperature, ΔT the temperature difference between the object
temperature and T_O, and σ Stefan - Boltzmann's constant.

It can be seen from Eq(1) that the sensitivity is improved by speeding up the scanning
velocity. If a constant line frequency is desired, the scanning width can be widened to
raise V_B, which can also be done by raising the line frequency(the number of lines will be
increased correspondingly), but the band width has to be wider and the noise level from
the preamplifier to be correspondingly higher. However, we usually adjust the scanning
raster to a place beyond the effective target area when PEV is in operation. If the field
frequency is lowered and the field period lengthened, more pyroelectric charges will be
accumulated, thus making the signal current stronger. Flicker can be eliminated only if
the storage is used, and the field frequency is changed to 50Hz through scan transformation
when reading. In practice, i_{sm} can not be inversely proportional to the field frequency
because of the influence of MTF. Calculations show that when the field frequency is lowered
by a factor of five the signal output will nearly double.

Fig. 1 shows the MTF curve relative to spatial frequency n and the chopping frequency
f_c. As is illustrated in Fig.1, the attenuation of MTF becomes faster with the increase of
spatial frequency. One way to solve the problem is to improve the target and the other to
introduce compensation in signal processing system by way of raising the high boost. It
should be noted that as the lateral heat diffusion will lower the resolution(it is similar
to the aperture effect), the phase - frequency characteristic must be kept linear to prevent distortion as the high boost is going up.

Because the output current of PEV is much less than that of the photoelectric one, the
output signal - noise ratio is rather low even if the band width of preamplifier is reduced to about 2 MHz, which brings about much difficulty for us to get a high boost. Such
a problem can be solved only by multiple-field cummulation processing. The value of the
chopping frequency f_c strongly influences MTF at a low spatial frequency. When a non - real
time storage (e.g. silicon target storage tube) is chosen for operation, the chopping frequency taken is usually lower so that a higher singal - noise ratio at lower spatial frequency can be obtained. But it must be higher than 12.5 Hz picking up mobile image and
adopting real time storage for image - difference processing(1,2,3).

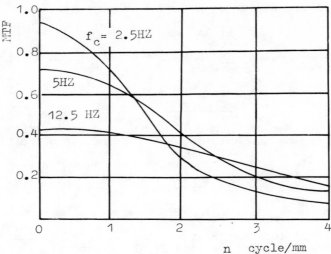

Fig. 1 MTF curve relative to n and f_c

The methods of the signal processing

In order to improve the sensitivity of PEV and overcome thermal spread effect of the target of PEV, several methods for processing signals from PEV are used in our infrared imaging system. They are the contour enhancement, pseudo color display and average of multiple frames and accumulation of multiple fields.

The vision is highly sensitive to the image contour, because it includes much important characteristic informations of the image. The experiment on visual psychology shows that image comprising the Mach overshoot edge is clearer than image of the practical edge. The contour enhancement is based on the feature of the psychophysics of the human visual system (4,5,7). In the method, the comb filter is applied, the horizontal and the vertical details can be extracted, the noise is subtracted then the details will be added to the original image, so that the contour of infrared image can be enhanced. The method can overcome thermal spread effect of the target of PEV (6).

The method of pseudo color display can stretch difference of gray degree between two adjacent pixels. Since the color is not real component, it is known as pseudo color. The color image gives human much more informations than the black and white image does. The human's eyes can watch more useful informations by changing the gray into the color. Our infrared imaging system has eight levels including black, blue, green, magenta, red, yellow, cyan and white. Analog signals instead of quantized ones entering into each level are transmitted so that more informations from infrared image could be picked up. Thus each of the colors corresponding to a special level is not uniform. Take yellow colour for example, its luminance varies from dark to light according to the amplitude of the signal entering into its level. Each color is a pure hue (i.e. 100% saturated color) although its amplitude (i.e. luminance) is varying. However any excessive change in luminance is not suitable because it is difficult to distinguish black from dusky. In the system, two principles in selecting a pseudo color coding should be emphasized. 1. a color between every two adjacent levels should be easily identified by eyes, 2. the coding sequence (from cold color to hot color) is to be in accordance with the temperature (from low to high).

When the image information is changed slowly, noise can be reduced, and resolution can be improved by averaging continuous multiple frames. It can be proved that the practical image signal is not changed and the radom noise power is down to 1/M by averaging M frames. With our system, the accumulation of multiple fields is used to improve the signal - noise ratio. Its principle is similar to average of multiple frames. In next section, the method is discussed in detail.

Multiple fields accumulation and image difference processing

The frame processor based on the storage tube is used for realizing signal processing of PEV, in our infrared imaging system. The frame processor has two functions: the increase of signal - noise ratio by multiple fields accumulation processing and the image difference processing for signal.

1. Multiple field accumulation

   Assume that the infrared thermal image picked up is stationary, the signals of each field are just the same, then the correlation coefficient is 1, and the correlation coefficient is zero for the random noises in each field are independent of each other, then after the accumulation of signals in a fields the signal-noise ratio of the output voltage becomes

$$\frac{S}{N} = 10 \lg n + 10 \lg \frac{e_S^2}{e_N^2} \text{ (db)}, \tag{2}$$

in which the second term on the right side is the signal - noise ratio of one field signal, and the first term is considered to be its increment. It is evident that this method can raise signal - noise ratio, e.g. accumulating ten fields of signals can increase the signal-noise ratio by as much as 10 db.

   Compared with five fields accumulation, we found by the monitor that twenty fields accumulation can greatly increase the image signal noise ratio (see Fig.2)

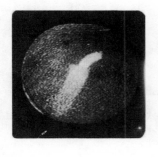   

            (a)   (b)

Fig.2   The infrared thermal image of an electric - iron

   (a) five fields accumulation   (b) twenty fields accumulation

2. Image difference processing

   By image difference processing we mean the process that the signal is delayed for half a chopper cycle, and then combined with the undelayed signal. So while the fixed noises in both open and closed fields and spurious signals formed by the imperfect compensation cancel each other, the useful signal increases by twice, thus greatly raising the image quality. It is very convenient to use storage tubes for image difference processing. We can write in the signals both in chopper open and closed fields one after another. They are first superposed each other on the target of the storage tube and then read out. Because of the non-linearity of the operating characteristic of the storage tube, the cancelling can not be done very completely, so a proper correction must be taken before writing.

   The block diagram of the frame processor is shown in Fig. 3. The voltage U_S coming from the signal shaper units corrected by γ corrector is fed to the strobing gate. When the storage tube is in a state of "writing in", the writing pulse U_W sent by logic control circuit makes the strobing gate open, and after being clamped, amplified and added to the composite blanking signal FS the signal is fed to the cathode of the storage tube. Then the video AGC will automatically keep the signal in the dynamic range. If the storage tube is in a state of "reading", the target switch is put on, and the reading signal being pre-amplified is fed to the shading compensation circuit. It is noted that when the storage tube is in operation, the beam and shading compensation must be corrected carefully so that the pedestal color can be adjusted to be as even as possible, otherwise it will make difficulties for the infrared imaging system to scale. As the reading signal will attenuate gradually with time, a video AGC is wanted to be connected to the output terminal. To sum up, the logic control circuit is to carry out the following operations: smearing, writing

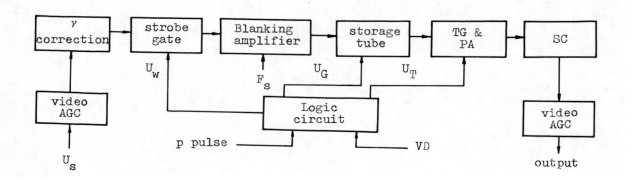

Fig. 3 Block diagram of the frame processor
TG & PA – Target Switch & Pre – amplifier, SC – shanding Compensation

and reading by "MAN" for the infrared image signal or TV test signal and those by "AUTO" at regular intervals.

Conclusion

The purpose of the signal processing is to raise the signal – noise ratio, lower the minimum resolvable temperature (MRT) and improve quality of the infrared image. A series of initiative experiments proved that the signal processing techniques discussed in the paper are of good performance. Thus the project is effective and feasible.

Acknowlegements

The author is much indebted to prof. Huang Tiexia (H.I.T., China), prof. Cai Danyu (H.I.T.) and prof. Zhang Shouyi (H.I.T.) for their valuable support and help.

References

1. Li Layuan, Huang Tiexia, Chinese Journal of Infrared Research, (1982), Vol, No.3, 161 – 170.
2. Huang Tiexia, Journal of HIT, (1980), 3.
3. Li Layuan, Review of the Academy of Electronic Infrormation Technology(1982, 4, 52 – 59.
4. Li Layuan, Telev. Eng., (1983), 3.
5. Li Layuan, Infrared and lasers Technoloty, (1981), 4, 11 – 18.
6. Li Layuan, proc. SPIE, 572 (1985).
7. Ketcham David I., proc. SPIE, 74 (1976).

INFRARED TECHNOLOGY XII

Volume 685

Session 2

Simulation, Modeling, and Testing

Chair
Randall Murphy
Air Force Geophysics Laboratory

Incorporation of angular emissivity effects in long wave infrared image models

John R. Schott

Center for Imaging Science
Rochester Institute of Technology
One Lomb Memorial Drive
P.O. Box 9887
Rochester, NY 14623-0887

Abstract

A radiometric model designed to facilitate target to background image modeling has been developed (c.f. Schott & Biegel 1985)[1]. The model was designed to incorporate atmospheric transmission, upwelled and downwelled path radiance and variations in these factors as a function of view angle. The radiometric model has been applied to kinetic temperature image models to simulate long wave infrared image characteristics, including spatial resolution and scale. A limitation of the model as presented to this point has been the lack of empirical data on emissivities of natural objects. This paper includes a description of two devices for measurement of the hemispheric and angular emissivity from large irregular surfaces. Emissivities for selected materials are presented and incorporated into the model. The effect of emissivity on simulated images is discussed. This model is being developed as part of an overall effort to expand long wave infrared image science. The potential for application of the expanded model to sensor performance evaluation and target recognition assessment are discussed.

Background

Schott 1982[2] defined a model for describing the long wave infrared (LWIR) radiance reaching an aerial or satellite sensing platform, as well as preliminary evaluation of certain features of the model. The model expresses the radiance at the sensor as:

$$L(h,\theta) = \tau(h,\theta)\,\varepsilon(\theta)\,L_T + \pi^{-1}F\tau(h,\theta)\iint r(\theta)\,L_d(\theta)\,\sin\theta\,d\theta\,d\phi$$

$$+ \tau(h,\theta)\,r(\theta)\,L_b(1-F) +$$

$$L_u(h,\theta) \tag{1}$$

where; $L(h,\theta)$ is the radiance reaching the sensor $wm^{-2}sr^{-1}$ at altitude h and look angle θ relative to the normal,

$\tau(h,\theta)$ is the atmospheric transmission from the earth surface to the sensor,

$\varepsilon(\theta) = 1-r(\theta)$ is the emissivity of the sensor,

$r(\theta)$ is the reflectivity of the sensor,

L_T is the blackbody radiance associated with a target at kinetic temperature T,

$L_d(\theta)$ is the downwelled sky radiance from the angle θ relative to the earth normal at the target,

ϕ is the azimuth angle of integration for the integration of $L_d(\theta)$ over the hemisphere above the target,

F is the fraction of the hemisphere above the target which is sky,

L_b is the radiance onto the target from non sky backgrounds.

$L_u(h,\theta)$ is the upwelled radiance from the air column between the target and the sensor. All the terms in equation 1 are wavelength dependent except F. For the examples discussed here the model was exercised over the 8-14 μm window.

The model requires target and background temperatures, target emissivity and the background shape factor as inputs. It uses the AFGL LOWTRAN code to compute atmospheric transmission values as a function of altitude and then computes upwelled and downwelled radiance from LOWTRAN (Kneizys, et al. 1980[3]) derived transmission values, radiosonde derived temperature profiles and an approach suggested by Ben-Shalom et al. 1980[4] to modify LOWTRAN. Schott and Biegel 1985 presented the results of synthetic image modeling using a version of this model. The model was simplified by assuming there were no reflected background effects and that all targets were either lambertian or specular reflectors (i.e. either $r(\theta) = r(0)$ for all θ or that only direct specular reflection of the appropriate region of the sky dome was occurring).

Under these assumptions, the third term on the right hand side of equation 1 vanishes and the second term becomes:

$$\frac{1}{\pi} \tau^{-1}(h,\theta) \; r(0) \int_{\phi=0}^{2\pi} \int_{\theta=0}^{\frac{\pi}{4}} L_D(\theta) \; \sin\theta \; d\theta d\phi$$

(2)

for lambertian surfaces or

$$\tau(h,\theta) \; r(\theta) \; L_D \; \theta$$

(3)

for specular surfaces

Synthetic scenes were generated of targets and backgrounds to which temperature and emissivities were assigned. The radiance of the scene into the direction of the sensor was then calculated and propagated to the sensor where it was convolved and sampled to simulate the sensor's instantaneous field of view and slant range. Using this approach, expected target to background signatures for an LWIR imager were simulated for various angular viewing conditions. A serious limitation to this approach was the lack of emissivity data for surfaces as they are commonly found (e.g. painted metals, beach sand, etc.), and in particular a lack of data on how emissivity varies as a function of look angle in the LWIR region. The next section describes two devices that were constructed to measure LWIR emissivities to overcome this limitation.

Emissivity measurement approach

Most devices for measurement of emissivity are designed to measure small (approximately 6 cm^2), reasonably smooth surfaces. These instruments in general are not readily adapted to measurement of the more spatially variant rougher surfaces observed with remote sensing systems (e.g. pavement, construction materials, natural surfaces, etc.). Several investigators (Buettner and Kern 1965[6], Lorenz 1966[8], Brown and Young 1975[7]) have developed procedures for measurement of more extended roughened surfaces. However, these procedures generally require extensive knowledge of surface temperatures, background temperatures, surround emissivities or other parameters that are difficult to measure and/or control and have, therefore, not been generally adopted. The two devices described in this section have been designed to minimize the number of variables that must be observed to reduce measurement errors.

The first instrument was designed to measure the normal hemispheric emissivity (actually reflective) of rough surfaces of approximately 10 cm. diameter. Figure 1 is a schematic of the apparatus. When the hot background is over the target, the radiometer successively views a low emissivity chopper blade, the sample and a high emissivity chopper blade. The cold background is then moved into place and three more readings of these same surfaces are made.

The radiance reaching the radiometer can be expressed as:

$$L(h,R_1) = \varepsilon_1 L_{T1} \tau + (1-\varepsilon_1) L_h \tau + L_a$$

(4)

$$L(h,R_2) = \varepsilon_2 L_{T2} \tau + (1-\varepsilon_2) L_h \tau + L_a$$

(5)

$$L(h,S) = \varepsilon_s L_{Ts} \tau + (1-\varepsilon_s) L_h \tau + L_a$$

(6)

$$L(c,R_1) = \varepsilon_1 L_{T1} \tau + (1-\varepsilon_1) L_c \tau + L_a$$

(7)

$$L(c,R_2) = \varepsilon_2 L_{T2} \tau + (1-\varepsilon_2) L_C \tau + L_a$$

(8)

$$L(c,S) = \varepsilon_s L_{Ts} \tau + (1-\varepsilon_3) L_c \tau + L_a$$

(9)

for each reference chopper blade and the sample with the hot background and each chopper blade and the sample with the cold background respectively.

where; $L(h, R_1)$ is the radiance at the detector associated with the hot (h) or cold (c) background and reference blade one (R_1), reference blade two (R_2) or the sample (S),

ε_1, ε_2 and ε_s are respectively the emissivities of each of the reference blades and the sample,

L_{T1}, L_{T2} and L_{Ts} are respectively the blackbody radiance values associated with the temperature of each reference blade and the sample.

τ is any transmission loss in the system,

L_h and L_c are the background radiance values from the hot and cold backgrounds respectively,

L_a is any additive target independent radiance reaching the sensor.

Equations 4, 6 and 9 can be recombined to yield:

$$G_{s/1} = \frac{L(h,s) - L(c,s)}{L(h,R_1) - L(c,R_1)} = \frac{\tau (1-\varepsilon_s)\left[L_h - L_c \right] = (1-\varepsilon_s)}{\tau (1-\varepsilon_1)\left[L_h - L_c \right] = (1-\varepsilon_1)} \tag{10}$$

which yields

$$\varepsilon_s = 1 - G_{s/1}(1-\varepsilon_1) \tag{11}$$

By adding in equations 5 and 8, a multiple point solution can be obtained for each measurement reducing measurement errors. The advantage of this approach is that the only variables on the right hand sides of equations 4 through 9 that must be known are the emissivities of the standard chopper blades, and in fact only one of these must be known. The combined signal differencing and ratioing tends to remove nearly all systematic noise effects. The primary assumptions made are that the temperatures of the reference blades, sample and background remain constant for the 30 - 90 seconds required for a reading. Reproducibility tests indicated that this assumption was quite valid for the conditions under which the measurements were taken. This instrument, when used with a single reference, had a demonstrated precision, of ±0.007 based on repeated measurements on several samples over a two week period (Dunn 1986[5]). The hot and cold background temperatures were at approximately 60 and 0 C respectively. The expected accuracy, based on error propagation, is approximately ±0.01 when a single standard reference of emissivity 0.94 ± 0.005 is used for samples with emissivities of approximately 0.9. The samples were all measured at ambient temperature (approximately 20°). Results should be slightly better for the dual reference approach described here.

The sensor used for these studies was a thermistor-bolometer filtered to a bandpass of 8-14 μm. Normal emissivity values, measured in this bandpass for several common construction materials, are listed in Table 1.

The device described above was not readily adapted for angular measurements. A second device which utilized the normal emissivity measurements was developed for angular measurements. A schematic of the instrument is shown in Figure 2. The instrument consists of a radiometer observing the sample at controlled angles while the sample is located in a cold background. The background is kept at a uniform cold temperature of approximately 10 C by circulating a mixture of methanol and water through the hohlraum. The mixture is chilled by circulation over dry ice (CO_2) in a secondary chamber. The sample, as in the case with the normal emissometer, is at approximately ambient temperature (24 C).

Each angular measurement consists of two radiometer readings. One viewing normally and one at the desired angle. The radiance at the two angles can be expressed as:

$$L(\theta) = \varepsilon(\theta) \tau L_T + 1 - \varepsilon(\theta)\tau L_{hT} \tag{12}$$

and

$$L(0) = \varepsilon(0) \tau L_T + 1 - \varepsilon(0)\tau L_{hT} \tag{13}$$

where; $L(\theta)$ is the radiance observed at angle θ from the normal to the surface,

$\varepsilon(\theta)$ is the emissivity at angle θ,

L_T is the blackbody radiance associated with an object at temperature T,

τ is any transmission loss in the system (τ should be 1 for the 0.3 meter path studied, we assume there is no target independent radiance reaching the sensor).

L_{hT} is the radiance from the hohlraum due to its temperature T.

The hohlraum is coated with a high emissivity paint, and its radiance is determined by converting the average temperature of several contact thermistors to blackbody equivalent radiance in the bandpass of interest. The thermistors are also used as feedback to adjust the flow to achieve a uniform temperature distribution. With the normal emissivity $\epsilon(0)$ value obtained from the first emissometer, all the items on the right hand side of equation 13 can be determined except for L_T, which is treated as a single term. Since the observed radiance L(0) is known, equation 13 can be solved for L_T. This value of L_T is then used in equation 12 leaving $\epsilon(\theta)$ as the only unknown. Equation 12 is solved for each value of $\epsilon(\theta)$ of interest. The radiance observed normal to the surface L(0) is checked after each measurement to insure that the temperature of the sample has not drifted and to recompute L_T if any observable drift has occurred.

Figure 3 shows the results of measurements on a relatively diffuse surface (concrete) and a specular surface (water). Because these emissivity values indicated a well behaved function, they could be utilized in the atmospheric propagation model by curve fitting and interpolation to obtain the emissivity to any angle. An error analysis on the results from the angular instrument is currently underway.

Model predictions

The emissivity values obtained from the instruments described above were utilized in the radiometric portion of the synthetic image model to predict the radiance reaching a sensor from a painted wood sample with a temperature of 288 K and a water sample with a temperature of 285.7 K for a particular set of atmospheric and viewing conditions characterized by a morning mid latitude summer atmosphere. The results of this analysis are plotted in Figure 4. The shape of the emissivity curve as a function of view angle for the painted wood is similar to that for concrete. A more clear representation of the phenomenon is shown in Figure 5, where the contrast between these two features expressed as apparent temperature difference at the sensor is plotted versus view angle. This figure represents what a daytime image of a painted wood pier with a water surround would look like. The pier would have an elevated temperature due to its lower thermal inertia. However, when viewed vertically, it would appear cooler due to its lower emissivity and the relatively cold sky conditions. As the look angle decreases, the emissivity of water decreases much more rapidly than for the wood to a null contrast condition at 60 degrees followed by a strong contrast reversal compared to the normal view. At very shallow view angles, the decrease in atmospheric transmission coupled with the lower emissivities for both samples tends to severely reduce contrast. This type of contrast reduction and reversal driven by a combination of target temperature, emissivity variation and atmospheric effects is common in this spectral region. These dramatic changes in image appearance can severely confuse image interpretation algorithms and interpreters.

In using the synthetic image model, the kinetic temperature and emissivity as a function of view angle are defined for each scene element. The apparent radiance at the sensor is then computed as defined in the Background section. The radiance field is then convolved and sampled depending on the projected instantaneous field of view (IFOV) of the detector element on the ground (or in this case onto the scene radiance model). The radiance at the sensor is then processed through a simple response function to yield a final simulated image. These final images can be used to more realistically simulate a particular imaging system or collection scenario against a target scene. In addition, the synthetic images can be used to evaluate the utility of target identification algorithms when atmospheric effects, look angle, sensor range, sensor response function, target temperature, background temperature and target angular emissivity are varied.

One of the most critical parameters in realistic simulation of scenes is the emissivity and particularly the angular emissivity of scene elements. The measurement technique described here provides an acceptable method of generating this parameter for surfaces commonly encountered. We should point out that the emissivities measured by this method are hemispheric - angular emissivities, not complete bi-directional reflectance functions (BDRF) and as a result do not completely account for all possible variations in reflected radiance induced by background variations. For most conditions we believe the hemispheric emissivities effectively emulate the observation conditions and that BDRF data could not be conveniently incorporated into the current model.

Table 1

Normal Emissivity of Building Materials
8-14 Microns

Material	Mean Emissivity
Wood Painted White	0.925
Wood Painted Black	0.905
Sand	0.947
Fiberglass	0.947
Old Galvanized Steel Painted	0.912
Old Galvanized Steel	0.937
New Galvanized Steel	0.202
Roof Shingle	0.945
Concrete Block	0.940

References

1.   Schott, J.R. and J.D. Biegel, "Kinetic Temperature Image Modeling from Thermal Infrared Satellite Images," presented to the SPIE 29th Annual International Technical Symposium on Optical and Electro-Optical Engineering, August 1985.

2.   Schott, J.R., "Target and Background Infrared Calculations for Tactical Space-Based Sensor Applications," prepared for Naval Research Laboratories, RIT Report #PSI 82/83-51-3, 1982.

3.   Kneizys et al., "LOWTRAN 5 Atmospheric Transmittance/Radiance Computer Code," U.S. Air Force Geophysics Laboratory, Report AFFDL-TR-80-0067, 1980.

4.   Ben-Shalom A., et al., "Sky Radiance at Wavelengths between 7 and 14 μm: Measurement Calculation and Comparison with LOWTRAN-4 Predictions:" Applied Optics, Vol. 19, No. 6, pp. 838-839, 1980.

5.   Dunn, M.J., "Effects of Look Angle on Emissivity of Common Roof Top Materials in the 8-14 m Thermography Band as Determined by a Novel Ratioing Technique," Masters Thesis, Rochester Institute of Technology, expected 1986.

6.   Buettner, K.J.K. and C.D. Kern, "The Determination of Infrared Emissivities of Terrestrial Surfaces," Journal of Geophysical Research, Vol. 70, No. 6, pp. 1329-1336, 1965.

7.   Brown, R.J. and G.B. Young, "Spectral Emission Signatures of Ambient Temperature Objects," Applied Optics, Vol. 14, No. 12, pp. 2927-2934, 1975.

8.   Lorenz, D., "Temperature Measurement of Natural Surfaces Using Infrared Radiometers," Applied Optics, Vol. 7, No. 9, pp. 1705-1710, 1968

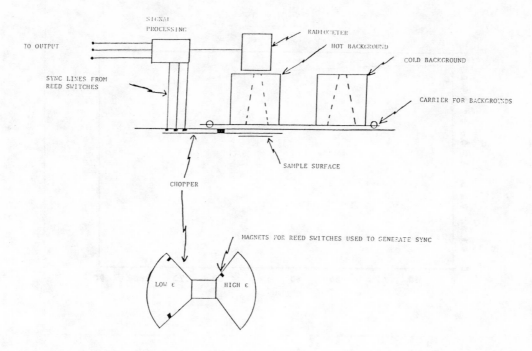

Figure 1 - Normal Emissometer Schematic. The chopper blade rotation provides consecutive radiance values for the two references and the sample as triggered by the reed switches. The background is then switched and the process repeated.

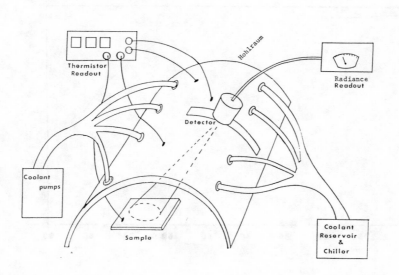

Figure 2 - Angular Emissometer Design. The thermistors on the underside of the hohlraum provide the background temperature and hence radiance. The radiometer views the sample first vertically and then at the angle of interest.

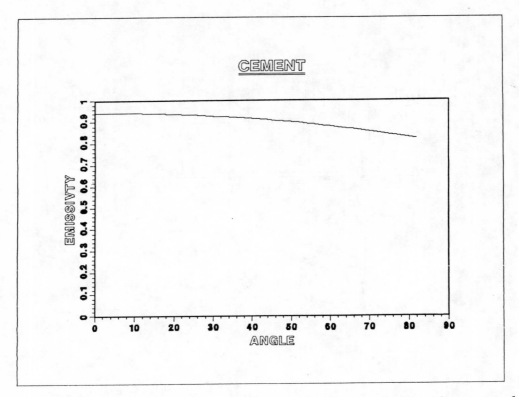

Figure 3a - Emissivities as a function of view angle for concrete. (note nearly lambertian surface behavior)

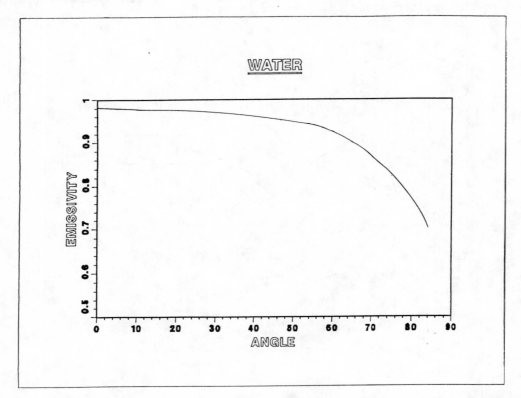

Figure 3b - Emissivity as a function of view angle for water. (note highly non lambertian behavior)

WOOD

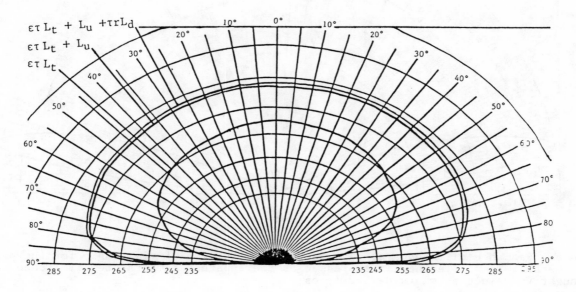

Figure 4a - Radiance at the sensor as a function of view angle for painted wood expressed as apparent temperature.

WATER

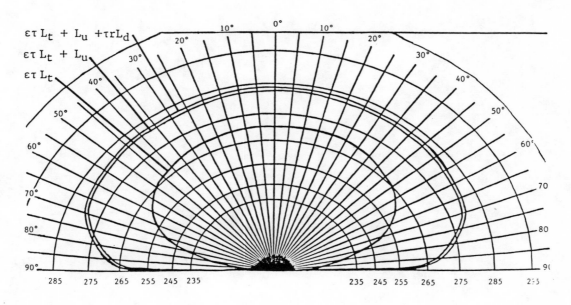

Figure 4b - Radiance at the sensor as a function of the view angle for painted wood expressed as apparent temperature.

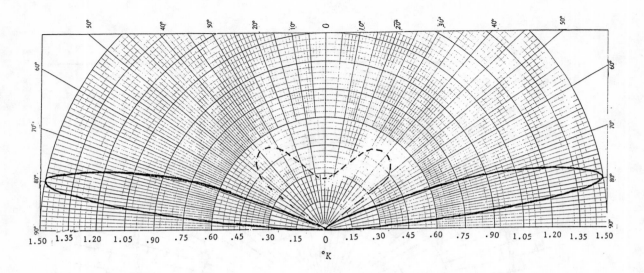

Figure 5 - Radiance difference at the sensor as a function of view angle for the painted
wood - water. The difference is expressed as the apparent temperature difference
with negative values shown as dashed lines.

AN INFRARED SCENE COMPOSER FOR ELECTRONIC VISION APPLICATIONS

Timothy G. Bates
Sandia National Laboratories
Division 324
P.O. Box 5800
Albuquerque, NM 87185

Michael K. Giles
Department of Electrical and Computer
New Mexico State University
Box 3-0
Las Cruces, NM 88003

Abstract

Because it is impossible to obtain real-image data that represent all of the scenarios and environmental conditions under which target recognizer models should be tested, we have developed a scene composer to aid in the analysis and evaluation of multiple-sensor electronic vision systems. The scene composer can present real-image data, computer-generated synthetic image data, and/or data composed of both real and synthetic imagery merged together so as to mimic an actual scene. For example, synthetic targets can be merged with real-image background scenes or vice versa. In addition, the composer uses a simple multiple scattering model to simulate image degradations due to fog or dust.

Presently, our synthetic generator is a simple infrared scene model based on the assumption that a given absolute temperature in a scene will be detected as a graybody radiance value. The temperature model allows wavelength-dependent functions such as surface emissivity, detector responsivity, and spectral filter characteristics to be included in the integration of the Planck equation. Real-image data have been obtained from the Texas Instruments 8-bit LANTIRN Database and from other sources.

The paper includes examples of real, synthetic, and merged infrared images and images degraded by simulated fog and dust.

Introduction

Due to excessive costs and limited accessibility, it is impossible to test and evaluate real, self-contained sensor systems in all possible scenarios and under all possible environmental conditions. A potential solution to this problem is the use of synthetic computer-generated image data in which targets are presented under various environmental conditions in backgrounds composed of differing features such as forests, deserts, grasslands, croplands, and waterways. Synthetic imagery allows for the merging of man-made features such as buildings, roads, and bridges into backgrounds in a manner such as to mimic an actual scene. Furthermore, it is possible to merge target images (real or synthetic) with background images (real or synthetic), to produce a very useful and flexible scene composer.

This paper describes a simple scene composer that is used to create input scenes for the evaluation of infrared sensor systems and automatic tracking algorithms. The scene composer consists of a basic scene model, an image degradation model, and target extraction and merging algorithms. The basic scene model, similar to that reported by Schott and Biegel[1], assumes that a given absolute temperature in a scene will be detected as a graybody radiance value. Wavelength-dependent functions such as detector responsivity, spectral filters, and surface emissivity are included in the model by successive multiplications before integration of the Planck function. The scene degradation model contains a point spread function model[2] that incorporates degradations in image quality caused by sensor and atmospheric parameters. The present model includes the effects of diffraction, aberrations, detector size, motion blur, background radiance, and multiple scattering by aerosols or dust. The multiple scattering model is based on an early theoretical result reported by Ishimaru[3]. The target extraction and merging algorithms used in this work are based on the pyramid technique developed by Burt[4]. Two versions of the scene composer are operational at this time. One is implemented on a VAX 11/780 using a Gould IP8500 image processor, and the other operates on an IBM PC using a Trowbridge PCIP100 image processor.

The following sections present more detailed descriptions of the scene and image degradation models followed by the results and conclusions. In the final section, example scenes are presented including a temperature scene and the radiance images which result when this scene is detected by sensors operating within two different wavelength regions, images degraded by fog, and a sequence of images that demonstrate the target extraction and merging algorithm. All of these images were derived from real-image data obtained from the Texas Instruments 8-bit LANTIRN database.

The Basic Scene Model

Images used as inputs to infrared sensor systems are composed of two-dimensional arrays of radiance pixels. The spectral radiance of a given picture element is related to its absolute temperature by the Planck function:

$$L(\lambda,T) = \frac{2hc^2 \; \varepsilon(\lambda)}{\lambda^5 (e^{-hc/\lambda kt} - 1)} \tag{1}$$

In equation (1), $\varepsilon(\lambda)$ is the emissivity function, T is the temperature in degrees Kelvin, λ is the wavelength of the radiant energy, h is Planck's constant, k is Boltzmann's constant, and c is the speed of light. The total effective radiance (i.e., the brightness) of each pixel seen by the sensor is obtained by integrating the Planck function over the spectral bandpass of the sensor system which is determined by the transmissivity $\tau(\lambda)$ of the optics and the responsivity $R(\lambda)$ of the detector. The transmissivity function is often defined by the spectral bandpass characteristic of a spectral filter placed in the system. Equation (2) is used to compute total effective radiance of a given pixel of absolute temperature T in degrees Kelvin.

$$L_T = \int_0^\infty \tau(\lambda) \; R(\lambda) \; L(\lambda,T) d\lambda \tag{2}$$

In practice, real-image data obtained from an infrared sensor of known transmissivity and responsivity are used to generate a temperature image by computing the inverse of equation (2) for every pixel element, or a synthetic temperature image is generated directly (computer-generated imagery). The resulting temperature image is then used to generate a simulated radiance image (by computing the radiance for every temperature pixel) approximating the image data that would have been obtained from another sensor having different transmissivity and responsivity within a totally different infrared spectral band. For example, real-image data obtained from a 3 to 5µm sensor can be used to generate a temperature image from which simulated image data from an 8 to 12µm sensor can be generated. This procedure assumes that the sensor that would have generated the simulated radiance image has the same F# and field of view as the sensor that originally generated the real-image data. The model is not restricted by this assumption, however, since the image degradation model can be used to account for the optical characteristics of the sensor being simulated (i.e., F#, detector size and field of view, aberrations, etc.). Examples of scenes generated in this manner are presented in Figure 1.

Image Degradation Models

Degrading effects introduced by the optics are incorporated in the simulated scenes by computing a system point spread function and then convolving it with the image data to produce the appropriate blurred image. The system point spread function is obtained by successive convolution of the spread functions due to each degrading effect (i.e., detector size, diffraction, motion blur, etc.). The pixel radiance values are scaled correctly by taking into account the peak of the responsivity curve and the radiant flux produced by each pixel radiance for a given sensor system using equations (3), (4), and (5).

$$s = R\phi \tag{3}$$

$$\phi = L_T A_d \Omega \tag{4}$$

$$\Omega = \frac{\pi}{4(F\#)^2} \tag{5}$$

s is the signal produced by the sensor (a voltage or current), ϕ is the radiant flux in watts, A is the detector area, and Ω is the solid angle subtended at the detector by the system exit pupil.

Image degradations due to propagation through random media such as fog or dust are incorporated in the scene composer by using Ishimaru's point spread function model. In this model, the point spread function is represented as the sum of "coherent" and "incoherent" spread functions, the "coherent" spread function being the spread of the unscattered light (the ordinary diffraction spread) and the "incoherent" spread function being the spread of the scattered light. These functions are presented in Chapter 14 of Ishimaru[3]. The model assumes plane wave propagation, large optical distance, a random medium consisting of uniformly distributed homogeneous spherical particles, particle size greater than the mean wavelength of the propagating infrared radiant energy, and a small scattering angle (forward scattering). Admittedly, these assumptions limit the general applicability of the model, but it is useful in generating qualitative image data that demonstrate the type of degradation one might expect due to fog or dust. Examples of images degraded by fog are presented in Figure 3.

Results

The scene composer has been used successfully to generate temperature images from input radiance images obtained from specific infrared sensors. Figure 1 shows a sample image obtained from a sensor in the 3 to 5µm region, the resulting temperature image obtained by transforming the input pixels, and a simulated 8 to 12µm image obtained by transforming the temperature image using equation (2). In this particular case, a flat spectral filter and unit emissivity were assumed for each transformation; however, the scene generation algorithm allows the user to input any desired spectral filter, emissivity and responsivity functions and then incorporates these functions in the integration of the Planck function as shown in equation (2).

Figure 1a. An Input IR Image Obtained Using
a 3 to 5μm Sensor

Figure 1b. The Temperature Image Obtained by
Transforming the Input Image of
Figure 1a

Figure 1c.   The Image of Figure 1a as Seen by a Simulated 8–12 μm Sensor

Figure 2 contains a sequence of images that demonstrates the image merging technique used by the scene
composer.   The truck in Figure 2a has been cut out and merged into Figure 2b.   The resulting merged image is
shown in Figure 2c.   The Burt pyramid method works well for this application.

Figure 2a. Original Infrared Truck Image

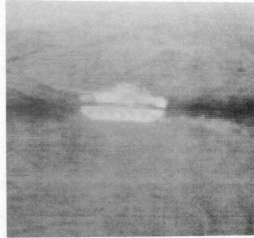

Figure 2b. Original Infrared Tank Image

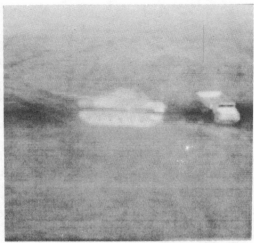

Figure 2c. Merged Image

Results obtained using the fog degradation option are presented in Figure 3. These images are useful as qualitative data only at this point, demonstrating that the most pronounced effect of fog is to reduce the image contrast significantly. The generation of quantitative degraded imagery that might be expected as a result of imaging through obscurants such as fog and dust will require the development of improved models and a significant data collection effort to validate those models.

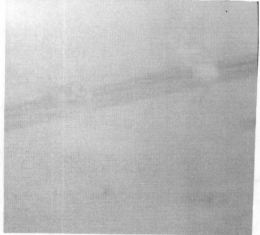

Figure 3a. The Truck Image of Figure 2a
with Simulated Fog

Figure 3b. The Tank Image of Figure 2b
with Simulated Fog

The scene composer is being used to compose scenes that are used as inputs to evaluate automatic target recognition and tracking algorithms. At the present time, libraries of targets and backgrounds are being assembled for use in this application. The scene composer has proven its usefulness in generating composite scenes from existing image data, in converting real-image data to temperature data, in converting image data taken with one system to approximately equivalent image data in a different spectral band, and in simulating image degradations produced by a variety of system parameters including scattering by random media.

References

1. Schott, John R. and Joseph D. Biegel, "Kinetic Temperature Image Modeling from Thermal Infrared Satellite Images," SPIE Proceedings No. 572, August 1985.

2. Bates, Timothy G. and Michael K. Giles, "Space-Variant PSF Model of a Spirally-Scanning IR System," to be published in SPIE Proceedings No. 685, August 1986.

3. Ishimaru, Akira, Wave Propagation and Scattering in Random Media, Academic Press, New York, 1978.

4. Burt, Peter J. and Edward H. Adelson, "Merging Images Through Pattern Decomposition," SPIE Proceedings No. 575, August 1985.

Maximum likelihood estimation of point source target amplitude and position in mismatched detector environment

L.R. Rochester

Ball Aerospace Systems Division
P.O. Box 1062, Boulder, Colorado 80306

Abstract

This paper describes the application of matched filter theory and maximum likelihood estimation to point source target position estimation in a scanning electro-optical system. The estimation is maximum likelihood at every pixel regardless of the noise and responsivity characteristics across the detector array. Algorithms are detailed and hardware is presented which is well matched to the implementation of the algorithms. Software issues are also described. The hardware is composed of a data flow processor and a general purpose signal processor, both manufactured by Signal Processing Systems, Inc.

Introduction

Maximum likelihood estimation is a powerful parameter estimation tool: the technique attempts to find the parameter value which is statistically most probable given a set of observed data. Maximum likelihood estimates have several desirable properties which make the maximum likelihood technique popular[1]. First, if an estimate exists which satisfies the Camer-Rao inequality, it is the maximum likelihood estimate. Second, the maximum likelihood estimate is unbiased. Third, the statistics on the estimate converge to a gaussian distribution as the number of observations approach infinity. Finally, if the parameters to be estimated have unknown probability distribution, the maximum likelihood estimate has lower variance than any other estimate.

This paper describes maximum likelihood estimation of the position of point source targets in a scanning electro-optical system. A particular signal processing scheme is presented. The scheme includes discrete matched filters to maximize the probability of target detection so we also relate the matched filters to maximum likelihood estimation.

Detector arrays often suffer from responsivity and noise differences from detector to detector. The algorithms presented here are designed to remain optimal over varying detector characteristics. A simple detector model is presented to support the description of filter coefficients.

Finally, we discuss a hardware and software architecture which implements the system. This architecture includes a data flow processor with dot product hardware, a general purpose signal processor with data flow software support, both commercially available.

The detector array

This section introduces a generic detector array. This array is used to pose the estimation problem and to demonstrate the signal processing implementation. The signal processor works for a much wider variety of detector arrays other than the generic array presented here. A brief description of how other arrays are used is discussed in a later section.

The generic detector array is shown in Figure 1. The target blur is scanned across the array horizontally. We call the horizontal direction in-scan direction. The dimension perpendicular to the in-scan direction is called the cross-scan direction. The objective of the signal processing is to estimate the cross-scan coordinate of the target peak and the time at which the target peak crossed an imaginary hairline as shown in Figure 1.

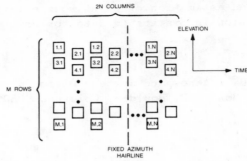

Figure 1. Generic detector array

To increase signal-to-noise ratio (SNR), the columns are time delayed and integrated (TDI). This is a common procedure in scanning electro-optical systems. When the detector noises are independent and identically distributed and the detector responsivities are equal, the SNR is improved by a factor of $\sqrt{N}$, where N is the number of detectors in the TDI sum. If the detector noise characteristics or the responsivities vary over the detectors in the TDI sum, then the integration of the TDI does not maximize SNR. In this case, each detector should receive a unique weight prior to integration: the weight should be a function of the detector's noise and responsivity characteristics. In a later section, we derive the optimum TDI weights, based on a simple detector model.

To support the ensuing analysis, we now digress to develop a detector model.

The detector model

We assume a detector model as shown in Figure 2. This model consists of photon energy falling on the detector aperture, the integration of the incident energy over the detector aperture, a multiplicative factor to model the detector's gain, an additive noise, electrical filtering, and temporal sampling. The responsivity and the additive noise are not necessarily the same for all detectors. This model also includes a background noise source which facilitates optimization of the signal processing to various background noise spectrums.

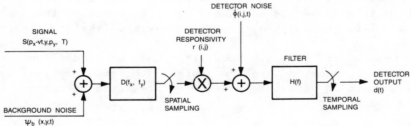

Figure 2. Detector model

To simplify later analysis, we develop expressions here for the signal and noise components out of the detector. We use the notation i and j to denote the i'th detector row and the j'th detector in the row on the detector array. Later, when we deal with the TDI'ed detectors, we use k to denote the k'th TDI output.

The signal model

Here we develop a mathematical representation for the signal into and out of the detector model. The input signal is the point spread function of a point source whose cross-scan position is ρ and whose peak crossing time is T. We seek to detect this target and to estimate ρ and T.

The input signal is represented as

$$\text{input signal} = s(x-x_o+vt, y-\rho) \tag{1}$$

where
       x = the in-scan coordinate on the focal plane
       y = the cross-scan coordinate on the focal plane
       x_o = the in-scan coordinate of the target peak without the scan
       ρ = the cross-scan coordinate of the target peak
       v = the scan velocity on the focal plane, and t is time.

The in-scan peak crossing time is related to x_o by $vT = x_o$.

The detector array is modeled by two steps. First, the detector aperture function is convolved with the signal, then the result is spatially sampled at the detector positions.

The detector aperture function is assumed to be

$$d(x,y) = \text{rect}(x,y) = \begin{cases} 1, & |x|, |y| < 1/2 \text{ detector width} \\ 0, & \text{otherwise} \end{cases} \tag{2}$$

The Fourier Transform of $d(x,y)$ is represented by $D(f_x, f_y)$.

The input to the spatial sampler is a convolution of the input with the detector function:

$$s^d(x-x_o+vt, y-\rho) = s(x-x_o+vt, y-\rho) * d(x,y) \tag{3}$$

The spatial sampler is represented as multiplication by a dirac delta function $\delta(x - x_{ij}, y - y_{ij})$ where x_{ij} is the in-scan position of the ij detector and y_{ij} is the cross-scan position.

The signal into the electrical filter is

$$s_{ij}(x-x_o+vt, y-\rho) = r_{ij}\left[s^d(x-x_o+ct, y-\rho) * d(x,y)\right]\delta(x-x_{ij}, y-y_{ij}), \tag{4}$$

where r_{ij} is the responsivity of the ij[th] detector shown in Figure 2. The electrical filter temporally filters the signal, i.e., the output of the filter is a temporal convolution

$$s^h_{ij}(x-x_o+vt, y-\rho) = s_{ij}(x-x_o+vt, y-\rho) * h(t). \tag{5}$$

Finally, the temporal sampler output is

$$s^*_{ij}(x-x_o+vt, y-\rho_o) = s^h_{ij}(x-x_o+vt, y-\rho)\sum_{m=\infty}^{\infty}\delta(t-mT)$$

$$= r_{ij}\left[s(x-x_o+vt, y-\rho)*d(x,y)\right]\delta(x-x_{ij}, y-y_{ij})*h(t)\sum_{m=\infty}^{\infty}\delta(t-mT) \tag{6}$$

Now we treat the noise terms. We may treat the signal and noise independently since the assumed detector model is linear.

Background noise out of the detector

The background noise is handled just like the signal. We represent the background noise by $\psi_b(x+vt, y, t)$. By analogy with the signal case, the output of the temporal sampler is

$$\psi^*_{ij}(x+vt, y) = \sum_{m=-\infty}^{\infty} r_{ij}\left[\psi_b(x+vt, y)*d(x,y)\right]\delta(x-x_{ij}, y-y_{ij})*h(t)\ \delta(t-mT) \tag{7}$$

$$= r_{ij}\delta(x-x_{ij}, y-y_{ij})\sum_{m=\infty}^{\infty}\left[\psi_b(x+vt, y)*d(x,y)\right]\delta(t-mT)*h(t)$$

Detector noise

As shown in Figure 2, the detector noise is added after the responsivity and prior to the electrical analog filter. The output of the temporal sampler is easily shown to be

$$\phi^*(i,j,t) = \sum_{m=-\infty}^{\infty} \left[\phi(i,j,t) * h(t)\right] \delta(t-mT). \qquad [8]$$

We now develop a simple expression for the composite signal out of a detector: the target signal plus the background noise plus the detector noise.

The composite detector output

The detector output is discrete due to the temporal impulse sampler. We are thus led to a simpler "discrete" notation where the target signal samples, the background noise samples, and the detector noise samples are represented as

target signal sample from detector $i,j = r_{ij} \, s(\rho,T,i,j,n)$
background noise sample from detector $i,j = r_{ij}\psi(i,j,n)$
detector noise sample from detector $i,j = \phi(i,j,n)$.

where $s(\rho,T,i,j,n)$ is the signal component out of a detector at time n without the responsivity term, and where $\psi(i,j,n)$ and $\phi(i,j,n)$ are similarly defined. We assume that r_{ij} and the noise statistics are known. We also assume the mathematical form of $s(\rho,T,i,j,n)$ is known, i.e., all is known except ρ and T.

The detector output is simply the sum of the three components above. We write this composite signal as

$$d(i,j,n) = r_{ij}\{s(\rho,T,i,j,n) + \psi(i,j,n)\} + \phi(i,j,n). \qquad [9]$$

Having established a simple representation of the detector output, we now proceed to describe the signal processing chain.

The signal processing chain

In this section, we introduce the functional signal processing chain. The chain consists of three functions:

1. A three-dimensional matched filter which consists of three independent matched filters:

   a) A weighted time delay and integration (WTDI) which maximizes the signal-to-noise ratio out of a TDI sum. Each component in the TDI has its own unique weight,

   b) A cross-scan filter which maximizes signal-to-noise ratio across post weighted TDI samples, and

   c) An in-scan filter which maximizes signal-to-noise ratio over time samples from the weighted TDI.

2. A target detection algorithm (not discussed further in this paper) which consists of a simple clustering algorithm applied to adaptive thresholded data, and

3. Target elevation position and crossing time estimation which determines target elevation position to fractions of a detector width and the azimuth hairline crossing time to fractions of a sample period.

A top level functional diagram of the processing chain is depicted in Figures 3A and 3B. We seek a maximum likelihood estimate of target cross-scan position and in-scan hairline

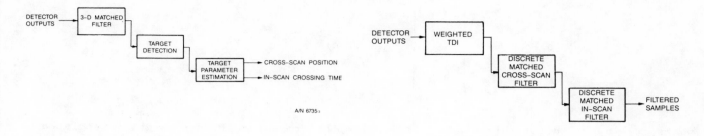

Figure 3A. Functional diagram of signal processing

Figure 3B. Expansion of 3-D matched filter

crossing time. Simultaneously, we seek to maximize the probability of target detection. It is well known that probability of detection is maximized by a matched filter receiver[2]. We thus want the processing up to target detection to constitute a matched filter.

We now discuss the mathematical details of the signal processing chain.

Mathematical details

Let us write the collection of time delayed detector outputs at time n as

$$\overline{D}(\rho,T,n) \overset{\Delta}{=} \begin{bmatrix} d(1,1) \\ d(1,2) \\ \vdots \\ d(1,N) \\ d(2,1) \\ \vdots \\ d(M,N) \end{bmatrix}_{\rho,T,n} \qquad [10]$$

We assume the time delays are precisely those needed to achieve the TDI. The weighted TDI transforms the detector output sample vector by

$$\overline{A}(\rho,T,n) = \begin{bmatrix} \overline{W}_1^T & & \\ & \overline{W}_2^T & 0 \\ & & \ddots & \\ 0 & & \overline{W}_M^T \end{bmatrix} \overline{D}(\rho,T,n) \qquad [11]$$

where $\overline{W}_i^T = \begin{bmatrix} w(i,1) & w(i,2) & \cdots & w(i,N) \end{bmatrix}$ = TDI weights for row i.

We note that the WTDI transforms MN detectors into M "effective" detectors.

If each detector row does not contain N detectors, then each TDI weight vector may vary in length, i.e.,

$$\overline{W}_i^T = \begin{bmatrix} w(i,1) & w(i,2) & \cdots & w(i,N_i) \end{bmatrix} \quad \text{where } N_i \text{ depends on the detector row i.} \qquad [12]$$

Alternatively zero weights may be used in entries associated with non-existent detectors. Hereafter we assume all rows contain N detectors. We are tacitly assuming the three dimensional matched filter is separable so that each dimension may be treated separately. We now consider the cross-scan and in-scan matched filters. The cross-scan matched filter performs the operation

$$\bar{B}(\rho,T,n) = \begin{bmatrix} \bar{h}_1^T \\ \bar{h}_1^T \\ \vdots \\ \bar{h}_M^T \end{bmatrix} \bar{A}(\rho,T,n) \qquad [13]$$

where $\bar{h}_i^T = \begin{bmatrix} h(i,1) & h(i,2) & \cdots & h(i,M) \end{bmatrix}$ = the cross-scan filter weights associated with TDI'd detector i.

The in-scan filter operation is

$$\bar{C}(p,T,n) = \text{Trace} \begin{bmatrix} \bar{g}_1^T \\ \bar{g}_2^T \\ \vdots \\ \bar{g}_M^T \end{bmatrix} \begin{bmatrix} \bar{B}(p,T,n) & \bar{B}(p,T,n-1) & \cdots & \bar{B}(p,T,n-q+1) \end{bmatrix}^T \qquad [14]$$

where $\bar{g}_i^T = \begin{bmatrix} g(i,1) & g(i,2) & \cdots & g(i,q) \end{bmatrix}$ = the in-scan filter weights associated with detector i.

This formula assumes the in-scan filter for each TDI'ed detector is distinct. Now we can rewrite transformations of the time delayed detector output data as

$$\bar{C}(\rho,T,n) = \text{trace} \begin{bmatrix} \bar{g}_1^T \\ \bar{g}_2^T \\ \vdots \\ \bar{g}_M^T \end{bmatrix} \begin{bmatrix} \bar{D}(\rho,T,n) & \cdots & \bar{D}(\rho,T,n-q+1) \end{bmatrix}^T \begin{bmatrix} \bar{h}_1 & \cdots & \bar{h}_M \end{bmatrix} \begin{bmatrix} \bar{W}_1 & \cdots & \bar{W}_M \end{bmatrix} = G^T D^T H W \quad [15]$$

where G,D,H, and W have the obvious meanings.

The last equation expresses the desired operation of pre-detection filtering. We desire that each g_i, h_i, and W_i to constitute discrete matched filters in their respective dimensions: in-scan time samples, cross-samples, and across detectors in the TDI.

<u>Determination of the filter coefficient</u>

The matched filter equation[3] is

$$\bar{c} = K^{-1} \bar{s} \qquad [16]$$

where K is the noise covariance of the input noise, $\bar{c}$ is the vector of filter weights to be determined, and s is the signal vector. We apply this equation to the determination of $\bar{g}_i$, $\bar{h}_i$, and $\bar{W}_i$. Consider the TDI weights: for signal input we have

$$\bar{S}_i = \begin{bmatrix} r_{i1}S \\ r_{i2}S \\ \vdots \\ r_{iN}S \end{bmatrix} \qquad [17]$$

where we assume that (1) the samples are already delayed by the amount necessary to implement the TDI, and that (2) the signal falling in a detector is a constant, S, after the TDI delay. The r_{ij} are the responsivities shown in Figure 1. The second assumption relieves us of having to use the general but cumbersome notation of equation 6. The noise covariance is given by

$$K = K_d + K_b \qquad [18]$$

where

K_d = the detector covariance matrix noise = diag $(\sigma_{ij}{}^2)$

and

K_b = background noise covariance matrix

$$= \{r_{ij}r_{ik}E[\psi(x_{ij} + vt_j,y)\ \psi(x_{ik} + vt_k,y)]\}_{j,k = 1,\dots,N} \qquad [19]$$

$$= \{r_{ij}r_{ik}R_b(x_{ij} - x_{ik},\ t_j - t_k)\}_{j,k = 1,2\dots,N} \qquad [20]$$

= {autocorrelation of the background noise evaluated for the spatial and time differences between detectors x the detector responsivities}

It is common to assume that the background noise varies spatially or temporally, but not both. The formula above models both variations.

Now the TDI weights are

$$\overline{W}_i = \begin{bmatrix} r_{i1}^2 R_b(0,0)+\sigma_{i1}^2 & r_{i1}r_{i2}R_b(\Delta_x,\Delta_t) & r_{iN}r_{i1}R_b((N-1)_x,(N-1)\Delta_t) \\ r_{i1}r_{i2}R_b(\Delta_x,\Delta_t) & r_{i2}^2 R_b(0,0)+\sigma_{i2}^2 & \vdots \\ \vdots & \ddots & \\ r_{i1}r_{iN}R_b((N-1)\Delta_x,(N-1)\Delta_t) & \cdots & r_{iN}^2 R_b(0,0)+\sigma_{iN}^2 \end{bmatrix}^{-1} \cdot \begin{bmatrix} r_{i1} \\ r_{i2} \\ \cdot \\ \cdot \\ r_{iN} \end{bmatrix} S \qquad [21]$$

where Δ_x = the spatial difference between the detectors and
Δ_t = the temporal difference (before time delaying) between detectors.

Now we consider some special cases.

Assume that there is no background noise for all of these cases. Thus the covariance matrix becomes

$$K = \begin{bmatrix} \sigma_{i1}^2 & & \\ & \sigma_{i2}^2 & 0 \\ 0 & & \ddots \\ & & \sigma_{iN}^2 \end{bmatrix} \qquad [22]$$

Special case #1

Assume that all detectors have equal responsivities and equal noise, except the first detector, which has unit responsivity but twice the noise ($\sigma_{i1}{}^2 = 2\sigma_{ij}{}^2$, $j > 1$). We have

$$\overline{W} = \begin{bmatrix} \frac{1}{2\sigma^2} & & 0 & \cdot \\ & \frac{1}{\sigma^2} & & \\ 0 & & \ddots & \\ & & & \frac{1}{\sigma^2} \end{bmatrix} \begin{bmatrix} 1 \\ 1 \\ \cdot \\ \cdot \\ 1 \end{bmatrix} S = \begin{bmatrix} \frac{1}{2} \\ 1 \\ 1 \\ \vdots \\ 1 \end{bmatrix} \frac{S}{\sigma^2}, \qquad [23]$$

where we see that the noisy detector gets half the weight of the others.

Special case #2

Let the first detector be dead: its responsivity is zero and noise power is infinite. Now

$$\overline{W} = \begin{bmatrix} \frac{1}{\infty} & & & 0 \\ & \frac{1}{\sigma^2} & & \\ & & \cdot \cdot & \\ 0 & & & \frac{1}{\sigma^2} \end{bmatrix} \begin{bmatrix} 0 \\ 1 \\ \vdots \\ 1 \end{bmatrix} \quad S \quad = \quad \begin{bmatrix} 0 \\ 1 \\ 1 \\ 1 \\ \vdots \\ 1 \end{bmatrix} \frac{S}{\sigma^2} \qquad [24]$$

Matching the cross-scan filter is also easy. Using equation 16 (the matched filter equation), we have

$$\overline{h}_i = K^{-1} \overline{S} \qquad [25]$$

where this time $K = K_d + K_b$ with

$$K_d = \text{diag} \left(\sum_j w^2(i,j)\sigma_{ij}^2 \right) = \text{post-TDI'ed detector noises} \qquad [26]$$

and

$$K_b = \left\{ \sum_j \sum_\ell r_{ij} r_{k\ell} w(i,j) w(k,\ell) R_b((j-\ell)\Delta_x + (j-\ell)\Delta_t, (i-k)\Delta_y) \right\}_{t,i,k} \qquad [27]$$

where $R_b(.,.)$ is the background noise autocorrelation out of detector. The signal vector is also more complicated:

$$\overline{S} = \left\{ \sum_j r_{ij} S(\rho,T,i,j,o) \right\}_i \qquad [28]$$

where we assume the signal is the detector model output signal evaluated at the in-scan peak. We associate the in-scan peak with time sample O without loss of generality.

The signal vector raises the issue of what value of ρ should be assumed when deriving filter coefficients. There are two approaches to this issue. The first approach is to have a number of ρ choices and corresponding supply of filters matched to each choice of ρ. Then the ρ is selected which maximizes the output given the observed data. This is a classical correlation receiver[2]. This approach, however, requires ever increasing computational resources as the number of post-TDI detectors grows and as the number of choices of ρ increases. To approach estimating ρ on a continuum, we must have a large number of ρ choices.

An alternative approach which is computationally less demanding with only a slight decrease in performance uses robust matched filtering[3]. The objectives of a robust matched filter is to maximize the worst case SNR. In our case, the worst case SNR occurs when the target is centered between two rows of detectors. Then once the target is located to the nearest detector row, a second filtering operation is performed to improve the estimate of ρ.

This filtering operation may be regarded as an interpolation[4] or a correlation receiver[2] as we shall see later. This analogy emphasizes the relationship between the mathematical problem presented here and standard signal processing techniques.

Given the large numbers of detectors on modern detector arrays and the high degree of position estimation accuracy, we will assume the latter approach is taken.

The in-scan coefficients also satisfy the general matched filter equation. Here the covariance is in the temporal dimension. We have

$$K_i = \{E[\Sigma w(i,j)(r_{ij}\psi(i,j,q) + \phi(i,j,q)) \cdot \Sigma w(i,k)(r_{ik}\psi(i,k,s) + \phi(i,k,s))]\}q,s=1,2\ldots,N_I$$

$$= \{\sum_j \sum_k w(i,j)w(i,k)r_{ij}r_{ik}(j-k, q-s) + \sum_j w^2(i,j)\sigma_{ij}^2 R_d (q-s)\}_{q,s} \qquad [29]$$

where we included the detector noise correlation introduced by the analog temporal filter of the detector model – this correlation is represented by $R_d(\cdot)$.

The signal vector is

$$\overline{S}_i(n) = \{\sum_j w(i,j)r_{ij}S(\rho,T,i,j,n)\} \qquad n=1, \ldots, N_I \qquad [30]$$

N_I = number of points in the in-scan impulse response. The same issue arises here with regard to our choice of T as with ρ in equation 28: again we elect to use the robust matched filter approach. Now we have seen how the filter coefficients are determined.

Now consider the last step – actual estimation of T and ρ. Assume that some unspecified target detection algorithm identifies the target from samples $\overline{C}(\rho,T,n)$ as given in equation 15.

Maximum likelihood estimation of T and ρ.

At this point in the signal processing chain, the target has been detected and a collection of data around the target peak is available for additional processing. Estimates of T and ρ are obtained from this data.

Intuitively, we have a rectangular matrix of samples around the peak and the algorithm will find T and ρ by correlating a pattern corresponding to specific values of T and ρ. The pattern with the highest correlation determines T and ρ. We show that this technique is maximum likelihood.

The input samples are an array of samples

$$V_i(\rho,T) = W_i[\overline{C}(\rho,T,n) \ \overline{C}(\rho,T,n) \ \ldots \ \overline{C}(\rho,T,n_p)] \qquad [31]$$

where $\overline{C}(\rho,T,n)$ is specified by equation 15 and W_i is a matrix of the form

$$W_i = \begin{bmatrix} 0 \ldots 0 \ 1 \ 0 & & \ldots & 0 \\ & 0 \ 1 & & \\ \vdots & & \ddots & \vdots \\ & & 0 & \\ 0 & & 0 \ 1 \ 0 \ldots 0 \end{bmatrix} \qquad [32]$$

which has n_p columns corresponding to the n_p $\overline{C}$-vectors in equation 31 and L rows where L is the number of rows used for parameter estimation. The diagonal of 1's is centered on the ith column, where detector i was identified by a detection algorithm as containing the peak sample. The value of L is selected based on the cross-scan dimension signal extent and the computational abilities of the signal processing hardware.

Let us now apply the maximum likelihood technique to the matrix $V_i(\rho,T)$. To do this, let us arrange the columns of $V_i(\rho,T)$ into one large vector

$$\overline{V}_i(\rho,T) \triangleq \begin{bmatrix} \overline{C}(\rho,T,n_i) \\ \overline{C}(\rho,T,n_2) \\ \vdots \\ \overline{C}(\rho,T,n_p) \end{bmatrix} = \begin{bmatrix} \overline{C}_1 \\ \overline{C}_2 \\ \vdots \\ \overline{C}_n \end{bmatrix} \qquad [33]$$

where we dropped the ρ,T notation to emphasize that this vector is simply a vector of observed data: ρ and T are to be determined from the data.

The maximum likelihood technique seeks to maximize the conditional probability density of the parameter vector (ρ,T) given the observed data[1]. This conditional probability density is called the likelihood function. Assuming gaussian probability densities, we have

$$L(\rho,T:\overline{V}_i) \triangleq P(\rho,T/\overline{V}_i) = G(\overline{m}(\rho,T),K_v) \qquad [34]$$

where $G(m,K)$ denotes a multivariate gaussian probability density with mean vector m and covariance matrix K. We seek to find the ρ and T which maximize $L(\rho,T:\overline{V}_i)$. As is commonly done, we maximize the log likelihood, i.e., $\log(L(\rho,T:\overline{V}_i)$. This results in

$$\underset{\rho,T}{\text{minimize}} \quad [\overline{V}_i - \overline{m}(\rho,T)]^T K_v^{-1}[\overline{V}_i - \overline{m}(\rho,T)]. \qquad [35]$$

Let us simplify the problem and find the $\overline{m}(\rho,T)$ which minimize the log-likelihood. The optimum solution[5] is

$$\overline{m}_o(\rho,T) = K_v^{-1}\overline{V}_i. \qquad [36]$$

We will use this result later. An alternative approach to achieving the same result is to minimize the quadratic equation in (35) directly. That is

$$\underset{\overline{m}(\rho,T)}{\text{minimize}} \quad [\overline{V}_i - \overline{m}(\rho,T)]^T K_v^{-1}[\overline{V}_i - \overline{m}(\rho,T)] \qquad [37]$$

which is equivalent to

$$\underset{\overline{m}(\rho,T)}{\text{minimize}} \quad \{-\overline{m}^T(\rho,T)K_v^{-1}\overline{V}_i + \overline{m}^T(\rho,T)K_v^{-1}\overline{m}(\rho,T)\}. \qquad [38]$$

The first term is simply a dot product of the vector $-\overline{m}^T(\rho,T)K_v^{-1}$ with $\overline{V}_i$. The second term is a constant depending on ρ and T. Thus we can achieve the solution to the maximum likelihood problem by performing the dot product $\overline{m}^T(\rho,T)K_v^{-1}(\rho,T)$ for all possible values of ρ and T and choosing the ρ and T which deliver the minimum. To do this on a computer would require infinite power since these are usually infinite ρ and T. Therefore, one must select a discrete set of ρ and T. This introduces errors if the true ρ and T are not in the set. Therefore, an engineering trade-off is required to determine how many ρ and T are needed to satisfy accuracy requirements. Alternatively, we can use equation 36 directly. However, the matrix calculation of equation 36 requires that we then solve for ρ and T given $\overline{m}_o(\rho,T)$. This is difficult if $\overline{m}(\rho,T)$ is a complicated function.

We note that as the number of ρ and T in the discrete approach converges toward infinity, the solution becomes the maximum likelihood estimate. We also note that the minimization problem in equation 36 is equivalent to maximum a posteriori receiver[2], when the a priori probabilities of each ρ and T are equally likely.

The method of performing the dot products of $-\overline{m}^T(\rho,T)K_v^{-1}$ with the input is obviously a matched filter since it satisfies the matched filter equation when the signal is assumed to equal $-\overline{m}(\rho,T)$. Thus we construct a bank of matched filters matched to $\overline{m}(\rho_i,T_j)$ for some set of $\{(\rho_i,T_j): i=1,2,\ldots N, j=1,2,\ldots,M\}$. The estimate of ρ and T is that ρ_i and T_j which deliver the minimum dot product.

We have created a maximum likelihood receiver from separate matched filters – the WTDI, the cross-scan filter, the in-scan filter, and the ρ and T estimation proper. Each of these operations is optimal at every detector since the detector responsivities and noises were included in the derivation of the filter weights.

With this mathematical background we can now discuss the implementation.

Implementation

A wide variety of commercially available computer products has been applied to signal processing applications. The efficacy of these products for a specific signal processing application is usually measured by the percentage of rate throughput achievable for that application and the level of effort required to achieve that throughput. These measures are primarily determined by the architectural limitations of the system. One architecture which is particularly well suited to real-time signal processing applications is the data flow architecture.[6]

The data flow architecture consists of a number of processing elements (PEs) linked together; the distinguishing feature of the architecture is that the PEs execute their tasks only when the required operands (data) are available and there is a place to put the output. The data flow concept is not restricted to hardware. Data flow languages are also gaining popularity in the signal processing discipline. These languages describe signal flow graphs made of nodes and links. The nodes represent numerical or logical functions; the links connect nodes.

Signal Processing Systems[7] manufactures a data flow processor and a general purpose signal processor. Both are complemented with a data flow language.

The data flow processor is called the SPS-1000. The PEs of this machine may be any external processor or input-output device (with the appropriate interface) or one of SPS' Fast Fourier Transform processors or dot product processors.

The general purpose signal processor is called the SPS-81. This machine consists of four (nearly) independent virtual input-output processors and a high speed arithmetic processor. This processor has a data flow language which allows the user to program the processor in a fashion similar to the SPS-1000. The SPS-81 can also be used as an SPS-1000 PE. We are now prepared to discuss a system composed of an SPS-1000 processor with one or more dot product processors and an SPS-81 which evaluate equations 15 and 35, as well as the target detection algorithm.

For comparative purposes, we quantify the SPS-1000 and SPS-81 performance in Table 1. This table gives approximate performance and price figures of merit: the intent is to help the reader determine the performance and price range associated with solving the signal processing problem presented here.

Table 1. Summary of SPS-1000 and SPS-81

	Operations/Second	1024-Point Complex Integer Weighted FFT	Price
SPS-1000	Up to 1 billion, depending on configuration	600 μsecs	100–500K
SPS-81	Approximately 20 million	3 milliseconds	85–150K

The specifications of these processors and their software are available from the manufacturer. We use these processors here simply as a vehicle to demonstrate how the ρ and T estimation problem may be solved in a data flow context.

Flow graphs

Data flow languages and processors are based on flow graph representations of problems. We use an example flow graph shown in Figure 4 to demonstrate problem solving with the data flow technique. The flow graph shown in Figure 4 is a representation of the signal processing chain presented in previous paragraphs.

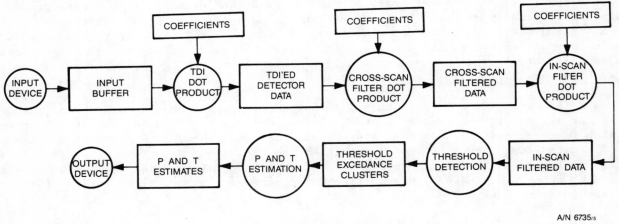

Figure 4. Flow graph

The flow graph in Figure 4 is composed of functions (circles), data buffers (rectangles), and paths (arrows) connecting functions and buffers. Many authors do not include buffers in the flow graph — the inclusion is standard practice with SPS programmers, since buffer definitions play an important role in the problem solution.

The data flow language developed by SPS requires that each buffer be defined in terms of dimensions. The paths describe how data is read or written from buffers: the access is specified in terms of the buffer dimensions. Using a derivative of the SPS data flow language, we will use the block diagram of Figure 4 to solve our problem. First we define buffers.

Let us assume that the data enters the machine from the input device according to this rule: for each sample time, read all the detectors in order ij for j=1, ..., N and i=1, ..., M with j the most rapidly varying. This suggests that the input buffer have dimensions of _I, _J and _Time. (We use the convention that mnemonics are preceded by an underscore.) Indeed, all the buffers admit very natural dimensioning as shown in Table 2. Although the buffers for our example have only two or three dimensions, the SPS software supports up to 256 dimensions of which up to seven can be used for any one buffer. The size of the buffers must also be specified, but we ignore that level of detail here.

Table 2. Buffer Definitions

Buffer Name	1st Dimension		2nd Dimension		3rd Dimension	
	Name	Range	Name	Range	Name	Range
Input	_T	1 to N	_I	1 to M	Time	Infinite
TDI'ed Detector Data	_Detector	1 to M	Time	Infinite	N/A	N/A
Cross-Scan Filtered Data	_Detector	1 to M	Time	Infinite	N/A	N/A
In-Scan Filtered Data	_Detector	1 to M	Time	Infinite	N/A	N/A
Threshold Exceedance	_IS Exceedance	1 to K	_XS Exceedances	1 to L	N/A	N/A
ρ and T Estimates	ρ	1 to 1	T	1 to 1	N/A	N/A

We must also define the paths. Paths are defined in terms of how the data is read or written in buffers. In the SPS data flow language, accesses are described in a manner similar to the way nested FORTRAN do loops are described. Typical descriptions are shown in Table 3.

Table 3. Typical Path Descriptions

Access Direction	Buffer	Access Description
Write	Input	_J = 1 to N _I = 1 to M Time forever
Read	Input	_J = 1 to N _I = 1 to M Time forever
Write	TDI'ed Detector Data	Detector = 1 to M Time forever
Read	TDI'ed Detector Data	(Detector = 1 to M) M times Time forever
Write	Cross-Scan Filtered Data	Detector = 1 to M Time forever
. . .	. . . etc.	. . .

Finally, the functions are defined. This requires that each function be associated with a PE, e.g., a dot product processor, an FFT, or an SPS-81 channel. The PE is instructed which path or paths to read and write. There are also a number of other parameters (most are PE-specific) which are not discussed here.

In the block diagram in Figure 4, we perform all the pre-target detection dot products in the SPS-1000; the remaining processing is performed in the SPS-81. This partitioning is sensible because (1) the target detection algorithm is well matched to the SPS-81 more general software and hardware capabilities and (2) the data rates after target detection are relatively low.

The ρ and T estimation is also a dot product operation (followed by a peak finder). The SPS-81 must execute this function whenever a threshold exceedance cluster is available.

Dot products and peak finding are part of the SPS-81 software library so that software implementation of this function is straightforward.

We see that implementing the block diagram of Figure 4 is simple with respect to the complexity of the algorithms involved. This implementation is optimal in the maximum likelihood sense for every detector (because it uses dot products rather than linear convolutions) regardless of the responsivity and noise characteristics of the detectors.

Special issues

The SPS-1000 uses complex arithmetic. This means that special care must be taken to process real data. When optimizing performance, the real data may be packed into complex words to achieve speed efficiency. We do not discuss this further here.

A host computer or bootstrap module is required to download programs and data to the SPS computers. In this paper, we ignored this issue. Like any other computer, the SPS equipment requires more programming effort as maximum throughput rates are approached. In the application considered here, the number of FFT PEs, the number of dot product PEs, and the amount of memory may also be adjusted to meet system requirements.

When the detector array has a geometry different than the generic array shown in Figure 1, the algorithms must change accordingly. In most cases this is easily handled by assuming each detector row has N detectors where N is the maximum number of detectors in any row. With this assumption, the data is addressed N samples at a time using TDI weights which are zero for detector samples not in the row. For example, suppose the array has a row of eight detectors followed by a row of four detectors followed by a row of eight followed by a row of four and so on. We assume a TDI length of eight and use eight weights on rows of eight. On rows of four, we use two detectors from the row of eight above and two from the row of eight below: the TDI weights apply a zero weight to the detectors from the rows of eight. This eliminates the need for variable length dot products and creates a simple data addressing sequence. Of course, it is also reasonable to perform all dot products with lengths equal to the corresponding detector row. This approach usually makes the software more complex.

Performance

Often the filter coefficient matrices are sparse, so that clever path specifications can eliminate unnecessary multiplications. In our case, pre-multiplication of H and W may reduce loading, depending on the dimensions of the respective matrices.

Another issue is latency. By changing the order of the operations and various path access schemes, the total processing latency can be modified. This is usually coupled with physical constraints such as loading and memory size.

Exact performance details of the SPS-1000 and SPS-81 are available from the manufacturer. However, in this section we use SPS timing figures to determine how many detectors we can process, using the generic detector array in Figure 1.

Let us assume that N=8 and determine M and the detector sample rate we can handle with the SPS-1000. A similar analysis is not presented for the SPS-81 since its loading depends on the complexity of the target detection algorithm and the number of targets. Let us assume that an 8-point dot product is used for the WTDI and cross-scan filter and a 16-point dot product is used for the in-scan filter. We will examine the cased when 1,2, and 3 dot product PEs are available. The number of dot products per time sample for each of the three SPS-1000 functions are: (1) M 8-point dot products for the TDI, (2) M 8-point dot products for the cross-scan filter, and (3) M 16-point dot products for the in-scan filter.

We can now plot M versus sample period for the case of 1,2, and 3 dot product PEs. We assume an 8-point complex dot product takes approximately six μseconds and a 16-point complex dot product takes approximately eleven μseconds. No provision is added to model system overhead: overhead times for the SPS-1000 are usually negligible. When one dot product PE is used, all filter times are additive. When two are used, we assume the 8-point dot products are performed in one PE; the 16-point dot products in the other. With three dot product PEs, each filter is performed on 1 dot product PE. In the case of multiple dot product PEs, we plot only the curve corresponding to the worst case PE.

With one dot product PE, the PE must compute M 8-point TDI dot products plus M 8-point cross-scan filter dot products plus M 16-point in-scan filter dot products all in one sample period. This means that

$$M \cdot 6 \ \mu secs + M \cdot 6 \ \mu secs + M \cdot 11 \ \mu secs \leq 1 \ sample \ period.$$

When two dot product PEs are used, the one which computes all 8-point dot products is the more heavily loaded. We have

$$M \cdot 6 \ \mu secs + M \cdot 6 \ \mu secs \leq 1 \ sample \ period.$$

In the case of three dot product PEs, the PE performing the 16-point PEs is most heavily loaded. Therefore we have

$$M \cdot 11 \ \mu sec \leq 1 \ sample \ period.$$

Our example is an illustration of the design issues associated with specifying the number of PEs. The case of three dot product PEs illustrates that adding PEs does not always improve performance significantly. In practice, it is desirable to load all PEs equally. This is accomplished by software and system level decisions regarding filter lengths.

The SPS-81 supports input and processing data rates for this type of application in the hundreds of kilohertz. In particular, the SPS-81 supports any data rate implied in Figure 5.

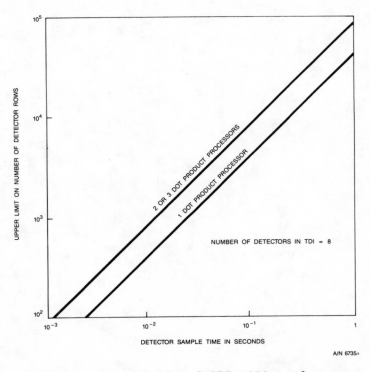

Figure 5. Example of SPS-1000 performance

If the real detector data is packed into complex words, these times can be significantly reduced (approximately by one-half). We do not describe this technique since it introduces several issues beyond the scope of this paper. There are also software techniques for more evenly distributing the computational loads over the PEs.

Summary

Given a detector array where each detector has its own responsivity and noise characteristics, we can obtain maximum likelihood estimates of the target position parameters using dot product processors and easily derivable filter coefficients. An implementation of such a system consists of an SPS-1000 data flow processor and an SPS-81 general purpose signal processor with data flow language support. Although the data flow architecture is not without problems, it is well suited to the problem discussed here.

References

1. Van Trees, H.L., Detection, Estimation, and Modulation Theory, Wiley, 1968.

2. Wozencraft, J.M., and Jacobs, I.M., Principles of Communication Engineering, Wiley, 1965.

3. Kassam, S.A., and Poor, H. Vincent, "Robust Techniques for Signal Processing", Proc. IEEE, Vol. 73, no. 3, March 1985.

4. Schafer, R.W., and Rabiner, L.R., "A Digital Signal Processing Approach to Interpolation", Detection of Signals in Noise, Academic Press, New York, 1971.

5. Whalen, A.D., Detection of Signals in Noise, Academic Press, New York, 1971.

6. Dennis, J.B., "Data Flow Supercomputers", IEEE Comput., Nov. 1980, pp. 48-56.

7. Signal Processing Systems, Inc., 223 Crescent Street, Waltham, Massachusetts.

Plume infrared signature measurements and comparison with a theoretical model

Santo Cogliandro, Paola Castelli

Infrared Group, I.A.M. Rinaldo Piaggio
Viale Rinaldo Piaggio, 17024 Finale Ligure (Sv), Italy

Abstract

Plume IR emission of turboshaft engines is investigated experimentally in 1-14.5 micron band by a Barnes Spectroradiometer with CVF and two IR scanners (AGA 782 Thermovision SW and LW) in order to compare it to a computer model and therefore to identify the most important parameters. This experimental set-up is completely adequate to field measurements of infrared signature from a turboshaft engine. It is completely interfaced to a HP 9836 computer to drive the instruments and to reduce on-line both spectral and spatial IR data. The spectral results are available as a function of wavelength (also in polar diagrams) and are connected to the emission wavelengths of the plume components. Spatial results are available in pictures (128x128 elements), displayed on a H.R. Colour Monitor. A sophisticated computer program allows to increase the detail of information. Spatial results are important to understand and to investigate phenomena like turbolence, non assial symmetric emission, etc. In order to estimate the real infrared signature of a turboshaft engine the transmittance of the atmosphere and the background emission have been measured during the field measurements. Also, absolute calibrations of spectroradiometer and scanners are conducted in our lab. Estimate of errors in terms of precision and accuracy of the measure are as well considered. Plume spatial infrared signatures of engine, for real installation on a rotorcraft, have been measured in order to adequate our computer model. This study is connected to a R&D program in the field of IR signature suppression for rotorcraft application.

Introduction

Today, the infrared energy emission (IR signature) investigation represents an important requirement in the construction of a military rotorcraft in order to estimate the potential survival against IR missile threat. Owing to the rapid development of IR seekers it is necessary, moreover, to consider and quantify means of infrared emission-reducing (IR suppression), through both theoretical evaluations and experimental measurements.

Theoretical assessments are necessary also for extrapolating results where measurement conditions are difficult or, even, impossible and researching the main physical phenomena involved. To this aim, we have developed a computer code describing IR signature of a rotorcraft with particular regard to the exhaust plume, because, as well-known, this can assure an important fraction of the total IR signature. To investigate phenomena useful to adjust our model (temperature distribution profile, turbolence, non assial-symmetric emission, etc) and to test it we have assembled a complete experimental set-up for target IR signature field measurements. A schematic view of the whole system is shown in figure 1.

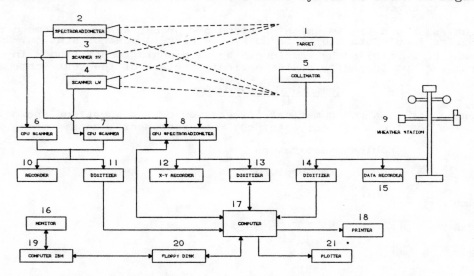

Figure 1. Target IR signature field measurement set-up

This set-up has been successfully used in IR signature measurements of two rotorcrafts, a Piaggio aircraft and a rotorcraft engine either "Un-suppressed" and "IR suppressed" by a mechanical device.

The study outlined in this paper deals with a rotorcraft engine plume IR radiation both theoretically and experimentally. Field measurements were performed at the Piaggio field engine test rig of Villanova airport. During the field measurements four types of data were recorded:

- ° Engine parameters to identify the test conditions
- ° Engine plume IR signature
- ° IR background emission
- ° Atmospheric transmittance

The field measurements of this rotorcraft engine meet the theoretical prediction.

Finally, we report spatial IR signature measurements recorded from a same type different mark engine installed on a military rotorcraft to recognize phenomena useful to improve our computer model.

The model

In the current literature several models have been used to describe IR signature of a rotorcraft. For plume IR emission prediction we have developped a model similar to that employed by Tracy-Jackson[1]. Spectral radiant intensity is calculated multiplying the Planck's black-body spectral emission function by the spectral emissivity and transmission of each homogeneous segment of the plume. We assign average plume properties for every segment while the transmission is calculated as a function of plume thickness and spectral location.

The total spectral radiant intensity I_λ emitted by the plume in direction of the aspect angle α is the sum of the plume segment spectral radiant intensity:

$$I_\lambda (\alpha , x_o , z_o) = \sum_i (I_\lambda)_i \tag{1}$$

where

$$(I_\lambda)_i = \frac{\varepsilon_i \, C_1 \, \lambda^{-5} \, \tau_i}{\pi(\exp (C_2 /\lambda T) - 1)} \tag{2}$$

x_o is the downstream position where the calculation is being made; z_o gives the radial position at which the calculation is being made at a particular downstream location. τ is the transmission along the chosen line of sight. ε_i is the spectral emissivity of the homogeneous segment of the plume. C_1 and C_2 are the Planck's law constants, λ is the wavelength and T is the plume segment temperature.

Very interesting is the 4.1 ÷ 4.8 micron band where CO_2 only is IR emitter. All other components of gas, including CO, have been included for the transmission calculations. The isoterm pattern in the plume has been tested by AGA thermovision scanner experimental data interpolation.

The atmospheric emission and transmission problems are processed by programs derived from the Lowtran 5 code[2]. The input data are recorded by the Piaggio weather station. Spectral radiant intensity must be multiplied by wavelength by wavelength transmittance in order to evaluate the spectral radiant intensity at the optical aperture of the test instruments.

Numerical technique

A Fortran IV program was used for numerical integration of the equations shown in the previous paragraph. The input quantities, obtained from experimental data, were:

T_{amb} = Ambient temperature

T_{gas} = Gas temperature

P_{amb} = Ambient pressure

P_{gas} = Gas pressure

v_{gas} = Gas velocity

A/C = Air-fuel ratio

CO_2, CO, O_2, N_2, H_2O = percentage of gas components

T_{ex} = nozzle exhaust radius

WAV   = wavenumber

Input data for Lowtran 5 transmittance code

The result furnished by the code represents the spectral radiant intensity as a function of wavelength at the optical aperture of the test instrument.

Infrared signature measurements

The engine under test is a Rolls-Royce GEM.2-MK.1001 delivering 750 HP (max continuous). The field engine test rig is designed in order to change both elevation and azimuth angles of sight.

During the field measurements engine parameters were recorded by the rig operator. The Piaggio mobile infrared van, electrically fed by means of a portable generator, was positioned at 200 meters from the engine in a plane normal to the jet axis. The elevation angle was maintained as small as possible. The hot engine sections were masked by a box at ambient temperature. The measurements were performed prior to sunrise to eliminate any direct solar reflection and to increase the target-to-background contrast. The instruments used were:

º Two calibrated IR cameras AGA 782 Thermovision SW (3.3 ÷ 5.5 micron) and LW (8 ÷ 12 micron) F.O.V. = 3.5 degrees.

º A calibrated spectroradiometer BARNES Mark II with a sandwich detector (InSb/HgCdTe) and a CVF covering 1.5 ÷ 14.5 micron. F.O.V. = 1 degree.

º A 10" collimator ELECTRO OPTICAL model 630 with reference source WS 153 (50 ÷ 1000ºC) for calibration and atmospheric transmittance measurements.

º A 12" extended calibrated source BARNES for calibration measurements.

º A weather station (air temperature and pressure, relative humidity, wind speed and direction).

Both IR cameras and spectroradiometer were mounted on a heavy tripod and placed inside a portable box for protection. The instrumentation signals were cabled to Piaggio mobile infrared van placed 20 meters behind the box.

An HP 9836 computer drove all the instruments and recorded the output signal on a floppy disk. Also, on request, it reduced on-line either spectral and spatial IR data.

Absolute calibration of spectroradiometer and IR scanners was conducted in our lab before and after each measurements. We used a collimator with a 1" calibrated black-body positioned at the minimum focal length of the device under test and at different apertures. The relationship between incident IR energy and output signal was parameterized by using the proper parameters of the instruments such as scanner F/n, Thermal Level, Thermal Range and Spectroradiometer output voltage, wavelength position, etc.). An HP 9836 computer drove the calibration procedure and recorded all the data on a floppy disk. During the field measurements, approximately every one and one-half hours the instruments were checked by means of an extended reference source positioned about 3 meters away, thereby ensuring that the source aperture overfilled the F.O.V. A significant effect on calibration in the 4 micron region was its performance in an atmospheric path, through which atmospheric transmission was assumed to be 100%. Actually, in this spectral region there was a remarkable CO_2 absorption but it was ignored. The error caused the overevaluation in the spectral signature and in the atmospheric transmittance at 4 micron wavelength band (5% approx.).

During the field measurements atmospheric transmittance was determined measuring the incident flux from a collimator with a calibrated known-temperature black-body at the same distance of the engine.

To obtain the elimination of the background radiation and its possible increase in temperature between two consecutive experiments, acquisitions before, after and during the tests were performed and recorded with the engine inoperative. The measured radiation should therefore have been that of the plume independently of its environment.

The precision of the measurement in term of RMS was about 2%. The infrared signature was evaluated with a 10% accuracy, owing to systematic errors like alignment, accuracy of range determination, radiometer optical focus errors, etc.

Data processing

To obtain spatial results from AGA Thermovision 782 the infrared images, once recorded on magnetic tape, were processed with proper computer programs employing the absolute calibration values previously obtained. Apparent temperature profiles may be obtained inverting Planck's formula for comparison with real time termistor measurements to evaluate the plume emissivity. We may represent IR images in different forms:

° symbolic representation employing up to ten alphanumeric symbols to show isothermal and/or isoradiant areas.

° coloured or grey representation: this gives an higher resolution of the punctual infor mation of the radiometer.

This resolution is up to 256 colour tones, enabling us to have more information on the behaviour of the emitted infrared radiation and/or the isothermal curves of some details. In order to visualize thermal images we use a coloured monitor raster-scan terminal. A sophisticated computer program designed and developed for Piaggio by CNUCE (an Institute of National Council of Research of Italy) allows to increase the detail of information up to a maximum of 1024 x 1024 pixels by using computer programs dealing with service mathemathycal, statistical and geometrical functions.

To obtain the spectral radiant intensity I_λ (dim.: W/sr/μ) at the spectroradiometer optical aperture we used the following relationship:

$$I_\lambda = (H_{T_\lambda} - N_{B_\lambda} \omega) \times d^2 \qquad (3)$$

where:

H_{T_λ} = spectral irradiance at the spectroradiometer aperture (W/cm^2/μ)

N_{B_λ} = spectral background radiance (W/cm^2/sr/μ)

ω = spectroradiometer F.O.V.

d = distance target/optical aperture (cm)

The spectral results are available as a function of wavelength. Through Simpson's integration rule, we may obtain the radiant intensity (W/sr) in any selected wavelength band.

Spectral transmittance τ_λ was obtained by the following relationship:

$$\tau_\lambda = \frac{V}{K_{dc} \, H_\lambda} \qquad (4)$$

where:

H_λ = theoretical incident flux obtained without the presence of atmospheric attenuation (W/cm^2/μ)

K_{dc} = equivalent radiometer response factor for a chopped point source (V/W/cm^2/μ)

V = signal producted at radiometer by source seen through atmosphere (V)

The spectral results are available as a function of wavelength. It was possible to calculate the mean transmittance in any selected wavelength band.

Comparison of theory and experiment

For this report we choosed the following conditions:

° Engine = Rolls-Royce GEM.2 - MK.1001

° SHP = 500 HP, M.C. and I.C. ratings at ISA, SL condition

° Distance = 200 meters

° Elevation = 0°

° Azimuth = 90° (tail-on = 0 degree)

° Wavelength band = 4.1 ÷ 4.8 μ (theory) - 1.5 ÷ 5 μ (experiment)

° Atmospheric conditions:

Temperature	= 308.16 °K
Pressure	= 14.6 PSI
Relative humidity	= 72%
Wind speed	= 1 m/s
Wind direction	= North
Time	= Prior to sunrise

Theoretical prediction are reported in figure 2 and figure 3. Experimetal results are reported in figure 4, figure 5, figure 6 and figure 7.

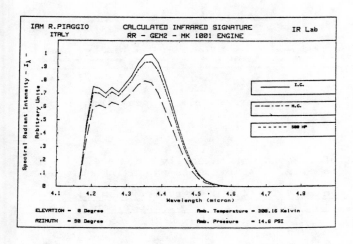

Figure 2. Theoretical prediction without atmospheric attenuation

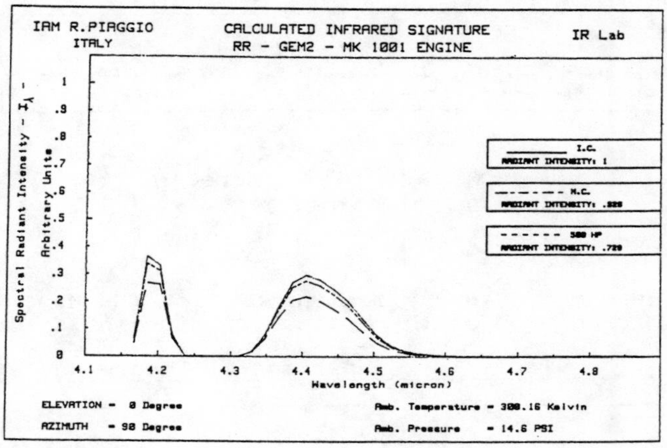

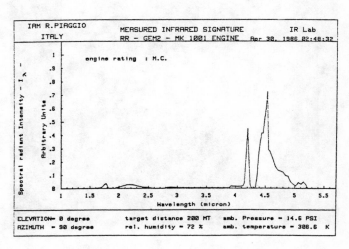

Figure 3. Theoretical prediction seen through atmosphere (distance 200 meters)

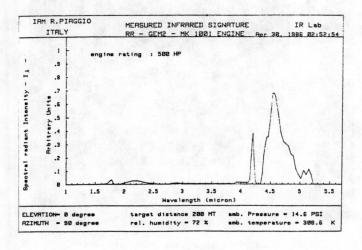

Figure 4. Experimental measurement. Engine rating: 500 HP

Figure 5. Experimental measurement. Engine rating: M.C.

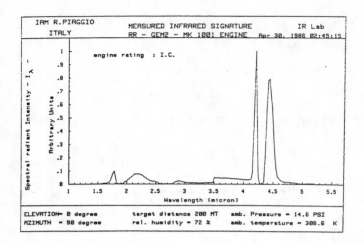

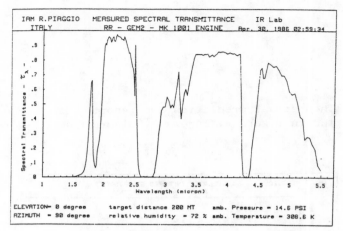

Figure 6. Experimental measurement
Engine rating: I.C.

Figure 7. Experimental spectral transmittance.

We compare theoretical prediction and experiment of the radiant intensity I in the 4.1 ÷ 4.8 micron band at the optical aperture of spectroradiometer. Comparison is shown in table 1.

Table 1. Comparison of theory and experiment $I(4.1 \div 4.8 \mu)$ Arbitrary units

No.	SHP	Theory	Experiments
1	I.C.	1	1.05
2	M.C.	0.926	0.88
3	500 HP	0.728	0.81

Infrared spatial results are reported in figure 8 (3 ÷ 5 μ) and figure 9 (8 ÷ 12 μ).

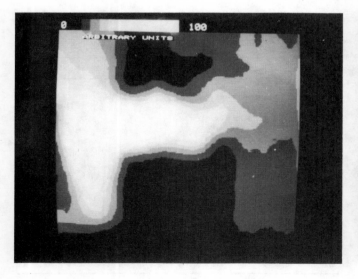

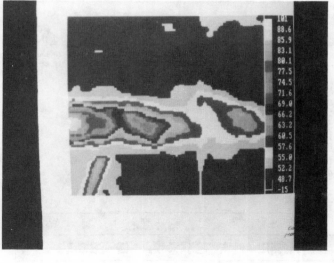

Figure 8. Infrared spatial measurement
(3 ÷ 5 micron)

Figure 9. Infrared spatial measurement
(8 ÷ 12 micron)

In figure 10 we report also the infrared spatial results in the 3 ÷ 5 μ band of a very similar engine installed on a military rotorcraft. We can note phenomena connected to the installation on a rotorcraft like rotor down-wash and warming-up of the tail.

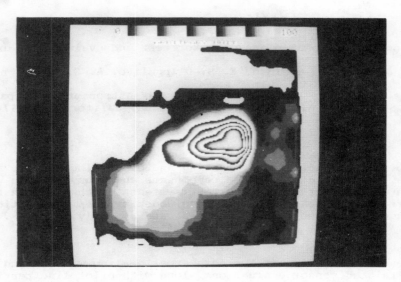

Figure 10. Infrared spatial measurement of an engine installed on a rotorcraft (3 ÷ 5 μ)

Conclusions

Our experimental set-up is completely adequate for field measurements of rotorcraft infrared signature in 1 to 14.5 micron band with a 10% accuracy. Though our computer model has given sufficiently good results it will be the subject of futher systematic studies by our team in order to understand and better quantify the phenomena connected to the engine installation on a rotorcraft. The techniques developped are applied to the problem of reducing the infrared signature of exhaust plume.

Acknowledgments

A particular thank is addressed to Mr. Marco Panizza (I.A.M. Rinaldo Piaggio Co.) for his help in computer elaboration and Miss Nadia Vivaldi for typing the manuscript.

References

1.   Tracy Jackson H., "An analytical model for predicting the radiation from jet plumes in the mid-infrared spectral region" Report No.RE-TR-70-7, U.S.Army Missile Command Redston Arsenal, Alabama 35809, 1970.
2.   Kneizys F.X. and others "Atmospheric Transmittance Radiance: Computer Code Lowtran 5" Report No.AFGL-TR-80-0067, Air Force Geophysics Laboratory Hanscom AFB, Massachusetts 01731, 1980.

A description of the focal plane/detector test and evaluation lab at MDAC-HB

D. D. Beebe, J. J. Lowe, C. Sheldon, E. S. D'Ippolito, A. G. Osler, and W. F. Morgan

Electro-Optics Lab, McDonnell Douglas Astronautics Company
5301 Bolsa Avenue, Huntington Beach, California 92647

Abstract

A description of a test facility for testing and evaluating visible and infrared (IR) focal plane arrays (FPA's) and associated components and subsystems is given. The facility is comprised of three computer controlled test systems for characterization of hybrid FPA's, detector arrays, and readout electronics under cryogenic conditions. Facility capabilities include FPA assembly and dewar test and assembly.

Introduction

This lab was developed to support business areas dealing with IR and visible sensors which include R&D related sensor evaluations along with pilot line testing. The suitability of detector arrays, readout devices, and FPA's for integration into sensor systems can be determined. Acceptance testing of delivered sensors can also be accomplished.

Capabilities exist to assemble detectors and readout devices into FPA's, to integrate FPA's into laboratory or tactical dewars, and to evaluate in-dewar FPA subsystems in terms of radiometric sensor characteristics and dewar performance.

Laboratory capabilities are depicted in Figure 1 in the form of a flow diagram. It illustrates the sequence in testing from FPA components to subsystems packaged in dewars to be inserted into a sensor system. It also shows our assembly, packaging, and integration capabilities.

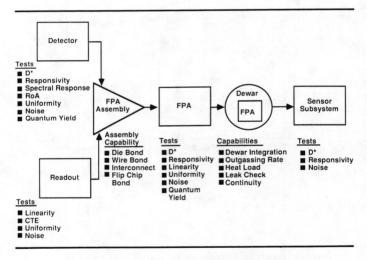

Figure 1.   Laboratory capabilities

The lab is contained in an 1800 square foot class 10,000 clean room with class 100 laminar flow benches. The walls, ceiling, and floor are metal and an elaborate grounding system is provided to reduce noise. To facilitate the test functions of Figure 1, the lab has three automated test systems to correspond to the detector, readout, and FPA subsystem levels of testing. Test automation is accomplished with three HP-1000 computers used for instrument control, test configuration, data acquisition, and data processing.

The detector and FPA test systems are interfaced to an in-house developed automated cryotest system. The readout test system is interfaced to a commercially purchased Flexion MP-3 manually operated cryotest system. The FPA test system also interfaces with an in-dewar test system for testing FPA's or detector arrays in dewars, or when testing FPA's in a subsystem.

Also included in the lab are assembly, packaging and alignment equipment. Devices, as well as assembled FPA's, can be die bonded and wire bonded in packages by die or flip chip bonding of detectors to readout devices. FPA's can also be aligned and installed into dewars and the dewar integrity can be verified.

Figure 2 shows the lab layout. Figure 3 shows the detector test system.

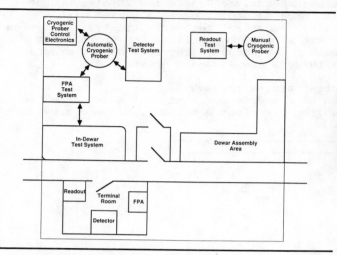

Figure 2.   Lab floor layout

Figure 3.   Detector test system

Electronic hardware

Room power

The clean room is powered by a dedicated 38 KVA motor-generator set. There is a 1" x 1/4" copper ground bus isolated from the metal walls at a four-feet high level around most of the perimeter of the room, terminated at a single point copper ground well (SPG). The room power outlets utilize isolated grounds insulated from the outlet boxes, conduit, and all building metal.

Computer systems

The laboratory equipment is controlled by three HP-1000 A900 computer systems. One computer controls the detector test system, one is dedicated to the readout test system, and the third controls the FPA test system, the in-dewar test system, and the automated cryotest system. The operating system is RTE-A (real time executive). The configuration of each computer system is listed in Table 1.

Table 1.   Computer Configuration

```
3 Megabyte RAM
132 Megabyte Hard Disk
HP-7970 Mag Tape Drive
HP-7914 Cartridge Tape
Two graphics terminals, one color and one monochrome
Five IEEE-488 I/O cards per system (each capable of controlling 14
       IEEE-488 compatible instruments)
Two 16-bit parallel I/O busses (capable of operating at 800 kHz)
One 16-bit buffered parallel I/O bus (capable of operating at 1.5 MHz
       continuous, or at a 3 MHz burst of 4K words)
One 6-pen plotter
One dot matrix line printer
Dual 710 KByte floppies
15 minute memory battery back-up
2400 baud rate modem
```

The line printers, plotters, and color terminals are in a remote terminal room, apart from the clean room. To maintain the low noise ground system, this remote equipment is linked to the computer systems via fiber-optic links.

Of particular interest in this configuration is the custom 1.5 MHz parallel I/O card. This card, together with our 14-bit plus sign Tustin A/D, enable a high speed, high resolution data acquisition capability.

Detector test system

The detector test system is configured to simultaneously characterize 20 elements of a detector array. To do this, it uses 20 HP-4140A pico ammeters, each in series with a separate Data Precision 8200 Calibration Source. For general test versatility, the system includes an HP-3497A Data Acquisition/Control Unit with 80 channels of low thermal EMF relays, a 5-1/2 digit DVM, and 16 general purpose 1 amp relays. There is also a programmable 10 amp, 0-60 volt power supply. All instruments are IEEE-488 controlled.

The system interfaces to the automated cryotest system through a five-foot umbilical. The current measurement accuracy and resolution of the detector test system are shown in Figure 4. These measurements were made through the five-foot umbilical. 1% accuracy down to 3.75 pico amp resolution has been obtained.

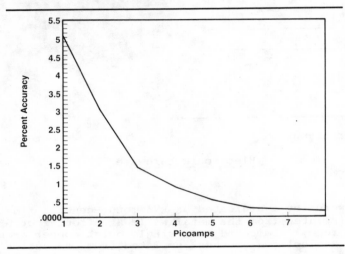

Figure 4. Current measurement resolution of detector test set

Readout, FPA, and in-dewar test systems

The readout and FPA test systems utilize separate but identical instrumentation racks. The FPA and in-dewar test systems utilize the same instrumentation rack.

The timing of each system is controlled by an HP-1000 programmable Interface Technology RS-4004 64-channel pattern generator with each channel being 4K deep. All 64 channels can run at 20 MHz, 12 channels will run at 100 MHz.

Interfacing to the RS-4004 are 16 Pulse Instruments CCD drivers capable of operating at 8 MHz with a +25 volt range, a 30 volt output swing, and programmable risetimes with a 2ns/volt to 2 microsecond/volt range. To provide bias levels, there are twelve 50ma +30 volt Pulse Instrument 702 bias supplies and six 150ma +30 volt current monitoring PI-720 bias supplies.

To minimize noise the Pulse Instruments equipment has had all AC power and fans removed and replaced with remote DC supplies and DC fans. The system uses an HP-3497A Data Acquisition and Control Unit with a 5-1/2 digit DVM to verify and monitor the levels of the pulse drivers and bias supplies through a coaxially switched calibration chassis.

Data is digitized using a Tustin 2015 14-bit bipolar A/D converter capable of operating at a 1.5 MHz sampling rate. The sample and hold setting time is 580ns to .003%. Figures 5 and 6 depict the computer/instrument/chamber interfaces. Figure 7 is a spectrum of the noise on the bias supplies in the MP-3 chamber at the probes. The peak noise is 30 μv.

General wafer probing

Using the cryogenic wafer probing chambers and an HP-4145A semiconductor parametric analyzer, characterizations of in-house designed wafers have been done to support our own development work in the area of CCD buffer amplifiers.

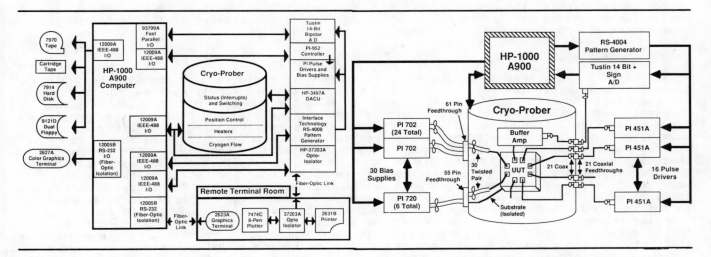

Figure 5. FPA and readout computer/
instrument interface

Figure 6. FPA and readout instrument/
chamber interconnect

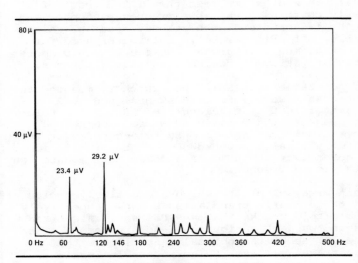

Figure 7. Bias supply noise spectrum
at probes

Cryotest systems

Manual cryotest system

This is a commercially procured FLEXION MP-3 manually manipulated cryogenic probing system designed to probe wafers, dice, or hybrid devices. The temperature range is 400K to 78K or lower with appropriate cryogens. The unit has various schemes for electrical access and illumination.

The test chamber consists of a thermally isolated, high vacuum assembly that incorporates the ability to mount, cool, illuminate, and probe the various devices under test. The test device is premounted on a removable cold chuck which is easily locked onto the chamber cold stage. An overhead cooled probe assembly accomplishes device probing. The chamber will accept a test device up to 2.0 inches square and 1.0 inch in height mounted on the chuck that facilitates either top or bottom illumination by externally mounted optical sources. The test device can be cooled to approximately 80K in less than 1 hour or 20K in less than 2 hours depending on device configuration and cryogen used. Temperature monitoring is provided by 10 platinum RTD's of which four are used to monitor/control chamber conditions and six are available for test specific purposes. Temperature set point and display are front panel and/or computer controlled. Chamber warm-up can be accomplished in less than 2 hours.

The device under test can be translated in X and Y directions in a plane normal to the optical axis by two externally mounted micrometers. Total translation in either X or Y direction is +1.0 inch with .0001 inch resolution. If probing of the test device is required, a cooled ceramic probe card up to three inches square can be configured which can probe four sides simultaneously. The probe card is actuated in the Z direction by an external micrometer with 0.0001 inch resolution. The entire probe mount can be rotated 360 degrees with micrometer control of +5 degrees. Electrical connections to the UUT include 21 coaxial feed-throughs and 30 twisted pairs providing access to the probes and a 55 pin feed-through to the bottom of the chamber for nonprobing applications such as substrate bias. Topside and bottomside illumination of the device passes through interchangeable, chamber mounted windows. If modulation and/or filtering of incident radiation is required, the sources are equipped with an optical chopper and 1.0 inch filter wheel. Available sources include blackbody and spectral standard visible sources. The test device can be viewed from the topside via a CCTV system that provides a full screen view of the probes in relation to the probe area of the test device. In addition to the visible CCTV system, a near IR camera system can be attached that allows viewing of the device to wavelengths of approximately 2.5 microns.

Automated cryotest system

The automated cryotest system consists of an MDAC-HB designed test chamber, test condition control electronics, and a continuous feed cryogen supply system. The system has been designed to provide a fully automated, computer controlled test capability for complete characterization of detector arrays, readout devices, and FPA's over a temperature range of 300K to 20K. Operation of all chamber functions, including temperature control, vacuum, UUT and probe translation, filter/optics positioning, and illumination is controlled and sequenced with menu driven software by an HP-1000 computer. Like the manual test chamber, the automated test chamber has a cooled, thermally isolated, high-vacuum mechanical assembly that provides the ability to mount, cool, probe and illuminate the unit under test. Illumination of the UUT can be accomplished in either flood, spot-scan or spectral illumination modes. Any of the illumination modes can be performed in topside or bottomside conditions by mounting the device, inverted or noninverted, in the cooled mounting chuck. Flood and spot-scan schemes utilize an internally mounted, re-entrant cone blackbody source with an operating range of 200K to 873K. The blackbody is mounted in the bottom of the chamber, facing upward through the source assembly, through the optical assembly, and onto the device and probe assembly. Spectral illumination is accomplished by a cooled, circular variable filter (CVF) which employs a tungsten ribbon filament lamp as its source, cooled optics and a precision, calibrated drive system to precisely select wavelengths from 1.5μm to 14.5μm. Typical half bandwidths are less than 1.5 percent of peak transmittances. The cooled source assembly contains a variable frequency chopper and an interchangeable aperture set to stop down the 1-inch diameter blackbody opening. Also included in the source assembly is a retractable, off-axis parabolic collimator used for spot scanning, and two coaxial filter wheels containing 6 interchangeable 2.0 inch diameter filters. All components in the source assembly are actively cooled with liquid cryogen.

The optical assembly consists of a cooled, rotatable turntable upon which is mounted the spot-scan assembly and CVF assembly. The spot-scan assembly is an MDAC designed, F/2 Schwarzchild telescope which images a 50μm diameter spot at wavelength of 10μm. The telescope is mounted on precision XYZ translation stages which provide 89 micro inch/step resolution over +0.6875 inches per axis. The maximum scan rate of the telescope, over this travel limit is 0.0893 inch/sec. The CVF assembly is also mounted on the XYZ translation stage so that its monochromatic image can be moved over the UUT. All components mounted on the optical turntable are likewise cooled with liquid cryogen.

The device and probe assemblies are two thin parallel plates which are actively cooled and contain an XY translation stage and a rotatable Z axis translation stage, respectively. They are designed to accommodate interchangeable cold chucks that the UUT and, if desired, probe cards configured for the UUT can be mounted on with various hold-down schemes. The XY translation stage allows positioning of the UUT about the optical centerline with 89 micro inch/step resolution and ± 0.625 inch travel per axis. The rotatable Z-axis translation stage allows the probes to be lifted from contact with the UUT. The translation covers 0.100 inches with 89 micro inch/step resolution, 0.015 inches of lift is controlled in an up/down manner for step and repeat probing; rotational translation allows probe alignment with UUT test pads. The cooled mounting chucks are made to accommodate devices up to 5.0 inches in diameter and 1.25 inches in height or probe cards up to 6.9 inches square. Probe cards can be fabricated to probe test pads 0.002 inches square minimum, spaced on centers of 0.0025 inches. By using the interchangeable features of the chucks virtually any combination of topside or bottomside illumination and probing schemes can be configured relative to the upward-looking source. A CCTV system allows viewing the UUT/probes from the top or bottom.

In-dewar test system (IDTS)

The ability to test and evaluate FPA's in either lab or tactical dewars is a critical step toward the integration of sensor systems. With the IDTS we are able to provide services ranging from basic acceptance tests to complete radiometric characterization for both visible and IR FPA's. Built with six axes of computer controlled motion and a custom dewar mounting bracket, the IDTS can handle many types, sizes, and shapes of either lab or tactical dewars. A 6000 psi dual gas system of Argon and Nitrogen is provided. The test capabilities of the IDTS fall into three categories: flood measurements, spectral response measurements, and spot scan measurements. The lab facilities for flood tests are as follows. The IR source is a CI SR-2-32 microprocessor based computer controlled blackbody. The SR-2-32 has a 1 inch cavity and features a temperature range of 50 to 1000°C with a temperature stability of ± .25°C and uniformity better than .5%. The aperture wheel has apertures ranging from 1.6mm to 22.2mm. The IR source can be chopped from 2.5 - 18,000 Hz. The visible flood unit utilizes interchangeable 45 or 200 watt (250-2500nm) standards of total and spectral irradiance with 0-5 kHz chopper. Aperture sizes and filters are flexible. Using the visible or IR flood sources coupled with the appropriate filters, spectral response studies from .4-14.5μm can be performed. The spot scan test capability is also available for either visible or IR FPA's. The visible spot scan test utilizes a quartz tungsten hologen lamp (nominal wavelength of .5μm), collimated by a precision off-axis parabolic mirror. It is focused to a 5μm diameter spot by a Nikon AI-S 200mm f4 lens mounted on a three axes computer controlled translational assembly (see Figure 8). With a linear resolution of .0001 inches and user friendly software, spot scan test can be efficiently performed. Also on this assembly is a Schwarzchild telescope capable of focusing an infrared glow bar with a nominal wavelength of 10μm to a spot diameter of 50μm. To further enhance spot scanning, the dewar mount assembly has three computer controlled orthogonal rotational stages, each with 5.0 arc minutes of resolution.

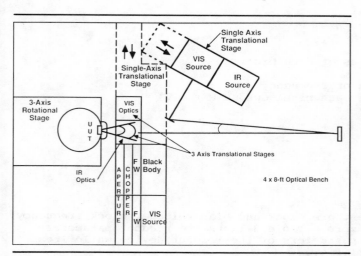

Figure 8.   Schematic representation of IDTS

Software Environment

Stimulation

The software to generate the UUT bias levels and timing waveforms has been written to facilitate the use of the Pulse Instruments bias supplies and the RS-4004 Pattern Generator which clocks the UUT through the Pulse Instruments pulse drivers. By being provided with a user friendly programming environment, the engineer is able to easily create a wide range of bias levels and timing patterns necessary to drive a variety of devices. The timing patterns generated in the high level environment of the HP/1000 are downloaded to the RS-4004 where they are output in real time to clock the UUT. This arrangement provides the user with a number of advantages over front panel operation of the RS-4004 by placing the user in the enhanced programming mode of the HP/1000 environment while still retaining the flexibility in timing control offered by the RS-4004. Operating the RS-4004 from the HP/1000, the programmer enjoys the use of a number of features not available from the RS-4004 front panel such as a "Logic Analyzer" mode which displays the generated timing patterns in graphical form for easy debugging. Other enhancements include extensive documentation, on-line help and error trapping while allowing the user to program in Word Generator, Timing Generator and Macro Program modes.

Device control and data acquisition

Once the correct bias levels and timing waveforms have been generated to operate the UUT properly the device is tested according to a user specified test plan in either the automated or manual test chambers. The controlling environment of the automated cryotest system permits the user to call up specific software routines that exist to run specific types of tests, such as FPA or detector array spot scan and spectral flood tests in a fully automated manner, including controlling probe position, device temperature, and chamber vacuum. By combining these prepared software routines the user can generate an automated test sequence. In addition to this test library, a high level programming environment has been established to assist the user in generating his own test procedures so he can easily effect a test not contained in the library. By combining the preprogrammed test sequences with user defined test procedures, test control becomes very flexible and easily tailored to a specific user application. After the test or test sequence has been defined, the HP/1000 controller runs the test and acquires the data according to user specifications. In the case of a detector array, data acquisition is accomplished through the pico-ammeters over the IEEE-488 bus. Pico-ammeter parameters in such a test are ammeter range, integration time and filtering. Data acquisition is accomplished using the high speed/high resolution A/D and the 16-bit parallel high speed I/O card in the HP/1000, both custom designs for this system. Input data rates of up to 1.5 megaword/sec with 14 bits plus sign resolution are possible in this mode.

Data reduction

Once the UUT has been sampled and the raw data has been collected, the data reduction process occurs. Given a set of test conditions, user specified output test parameters are calculated automatically. Each type of device tested will require a different set of performance parameters that will fully characterize the device. In a test of a detector array, the user variable test conditions are device bias, device temperature, device illumination, and array dimensions. The system can handle, but is not limited to, 512 x 512 arrays. Table 2 lists the detector parameters that can be determined. Items 1 through 12 are accomplished using a broadband flood test, items 13 and 14 require the spot scan test, and item 15 uses the spectral flood test.

Table 2. Detector Test Parameters

1. R_oA
2. σR_oA(Array)
3. Diode Curves
4. Responsivity
5. Array Average R and σR
6. Array Output Uniformity ($\sigma(V)$)
7. Array Noise Uniformity
8. Pixel noise N
   1) Function of frequency
   2) $\sigma(V)$ using multiple samples of each pixel
9. D*
   1) Using noise as a function of frequency
   2) Using multiple samples of each pixel output $N = \sigma(V)$
10. Array average D* and $\sigma(D*)$
11. Dynamic range
12. Linearity
13. Crosstalk (spot scan)
14. Pixel profile (spot scan)
15. Spectral Response
16. NEP
17. NEFD

The test conditions in a readout device test are clock and bias voltage, clock frequency, integration time, device temperature and fat zero. Table 3 lists the readout parameters determined in a test. Items 1 through 7 are parameters of the readout device while items 8 and 9 are specific to the device output amplifier.

Table 3. Readout Test Parameters

1. Charge transfer efficiency
2. Uniformity
3. Linearity
4. Dynamic range
5. Noise; ($\sigma(V)$) or function of frequency
6. Array noise uniformity
7. Charge capacity
8. Output conversion gain
9. Power dissipation

Tests of FPA's combine most of the variable test conditions and output device parameters of the detector array and readout device tests. The test conditions are device illumination, device temperature, device bias, integration time, and clocking frequency. Table 4 summarizes the parameters determined in an FPA test. Parameters such as RoA and diode curves are specific to the detector array and are not determined in this type of test.

Table 4. FPA Test Parameters

1. Responsivity
2. Array average R and $\sigma(R)$
3. Array output uniformity ($\sigma(V)$)
4. Array noise uniformity
5. Pixel noise
   1) Function of frequency
   2) $\sigma(V)$ using multiple samples of each pixel output
6. D*
   1) Function of frequency
   2) Using multiple samples of each pixel output $N = \sigma(V)$
7. Array Average D* and $\sigma(D*)$
8. Dynamic Range
9. Linearity
10. Saturation level/charge capacity
11. Power dissipation

Table 4 (continued)
12. CTE
13. OCG
14. Crosstalk (spot scan)
15. Pixel profile (spot scan)
16. Spectral Response
17. NEP
18. NEFD

Data presentation

Three types of data displays are available to the user: tables, histograms and two-dimensional plots. Tables of data can be listed as a given result for a user specified set of test conditions. Histograms showing the spread of the data points of a result can be generated for any device type and any set of test conditions. Figure 9 shows the dynamic resistance of a 2 x 16 HgCdTe detector array. Finally, two-dimensional plots of a test result versus a given test condition can be produced. This type of plot is very flexible, allowing the user to specify subsets of data and to display more than one curve on the same graph. Figures 10 and 11 show sample plots of visible CCD data and IR FPA data obtained on our systems.

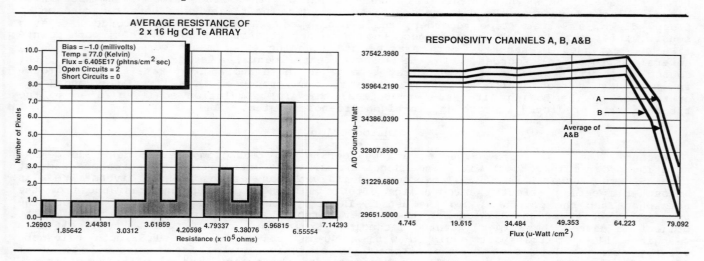

Figure 9. Dynamic resistance of 2 x 16 HgCdTe array

Figure 10. Sample plot of visible FPA data

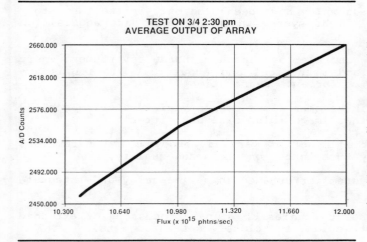

Figure 11. Sample plot of IR FPA data

Future upgrades

Planned improvements in the lab include modification of the detector test system which will enable automated AC characterization of detector arrays using a chopped source of illumination, enabling the calculation of D* as a function of frequency. To increase the data acquisition speed, 40 MHz pulse drivers, a high speed Data Ram 32 megabyte solid-state disk capable of operating at a 50 MHz rate, and a compatible faster 12-bit A/D will be acquired. A fiber-optic link interfaced to an Adage Video processor for real time video display and signal processing will be built. The Adage also can transfer data to a VAX and a Star array processor for accommodating larger FPA's faster.

Invited Paper

Test of IR arrays on the Kuiper Airborne Observatory

R. W. Russell, G. S. Rossano, D. K. Lynch, G. T. Colon-Bonet, J. A. Hackwell, T. C. Morse, R. H. Macklin, D. Murray, D. A. Retig, C. J. Rice, D. A. Roux, and R. M. Young

Space Sciences Laboratory, The Aerospace Corporation, P. O. Box 92957, Los Angeles, California 90009

Abstract

The Aerospace Corporation has conducted 5 flight series on the Kuiper Airborne Observatory (KAO) utilizing two-dimensional array infrared cameras. The KAO is operated by NASA Ames Research Center. These flights have several objectives with the primary task being the test in an aircraft environment of state-of-the-art two-dimensional arrays of both blocked impurity band and bulk silicon devices provided by Rockwell International and the Aerojet ElectroSystems Company under contract to the U.S. Army. There is extremely low crosstalk due to the SWIFET (switched FET) readout scheme employed on all three arrays flown to date. The ~ 500-element arrays were operated with a frame time of 307 μsec which placed a large burden on the data recording equipment. The flight data were recorded on 8" floppy disks, a 9-track tape recorder, and a high-speed 28-track flight recorder. System sensitivities have been shown to be good enough to detect astronomical sources, both point-like and extended in nature. Both staring and scanning experiments were performed using several astronomical sources. In the scanning experiments, a star was used as a point source which was scanned across the array at a fixed angle and angular rate using the rocking secondary mirror of the KAO. The data were then processed in a Time Delay and Integrate mode by the Boeing Aircraft Corporation; this TDI simulation achieved close to the theoretically predicted improvement in signal-to-noise ratio.

Introduction

Recently the U.S. Army engaged in an engineering and technology development and demonstration program in concert with wind tunnel and airborne astronomical research programs at the NASA Ames Research Center. The program deals with many issues in such diverse areas as atmospheric seeing effects, aerodynamics effects on images, background radiance from the atmosphere and the platform, and the use of IR arrays on airborne platforms. Because the program is built around the use of the Kuiper Airborne Observatory (KAO) (Figure 1), the project is dubbed the Kuiper Infrared Technology Experiment, or KITE. The KAO itself is a 4-engine C141 jet transport equipped with a 90-cm, all-reflective altazimuth telescope viewing out the left side of the aircraft at elevation angles from about 35° to about 72°. One major thrust of the KITE program includes the integration of existing state-of-the-art infrared detector arrays both of bulk silicon and of Impurity Band Conduction (IBC) architecture into flight-qualified dewars for testing under a variety of conditions on three different telescopes. The goals of the array work are:

1. To take several state-of-the-art IR arrays which have been used only in the laboratory and demonstrate their use in a working flight system which includes data acquisition and subsequent data processing.

2. To collect information on the use of the above arrays in an airborne environment where ground currents, radio transmissions, and mechanical and acoustical vibrations are only a few of many potential sources of performance degradation.

3. To quantify the angular size of the IR image near 11 μm over a variety of time scales, and to measure the amount of motion of that image (blur size and jitter).

4. To collect a database for evaluating the impact of varying aerodynamical parameters (for example, the mach number and cavity turbulence) on the image quality.

5. To attempt an actual demonstration of system performance in a time-delay-and-integrate (TDI) mode.

6. To provide a database for subsequent use in studies of signal processing of extended sources, background variability and subtraction, and closely spaced objects.

All of these goals were met in a 2-year time period, and within 5 months of the ambitious schedule conceived in August 1983, although we have only just begun to process the data and to interpret the results.

A project of this magnitude requires a tremendous amount of manpower. Organizing the effort was complicated in this case by the compressed schedule caused by funding delays and

KUIPER AIRBORNE OBSERVATORY

OPERATING CEILING:	41-45 K ft
ENDURANCE FOR RESEARCH:	7.5 hr
RANGE:	3300 n. mi
PAYLOAD:	60-100 K lb
TELESCOPE APERTURE:	36 in (91 cm)
POINTING PRECISION:	≤1 arcsec
INVESTIGATOR ACCOMMODATION:	5-7

LEAR JET OBSERVATORY

OPERATING CEILING:	45 K ft
ENDURANCE FOR RESEARCH:	2.5 hr
RANGE:	1300 n. mi
PAYLOAD:	1.2 K lb
TELESCOPE APERTURE:	12 in (30 cm)
POINTING PRECISION:	15 arcsec
INVESTIGATOR ACCOMMODATION:	2

Figure 1. Description of NASA aircraft.

aircraft problems. Figure 2 shows the organization of the project. We emphasize that this effort was conceived as a systems engineering project and not as a comparative array test and evaluation program. The three different arrays used were operated with different filters under different conditions; no quick comparisons of relative performance can be or should be made. Aerojet ElectroSystems Company (AESC) supplied a Si:Bi bulk array and an MC^2 array, both in a 16 x 32 pixel format, together with a signal processor which could be programmed to do a hardware flat-fielding to correct for offset and gain variations in the array/multiplexer combination. Because the same electronics had to be used for all devices this latter feature was used only in the laboratory. Rockwell International (RI) provided a BIBIB array in a 10 x 50 pixel format together with a signal processor built to identical interface specifications. This meant that the readout devices of the Boeing Aerospace Corporation and The Aerospace Corporation could connect to either set of electronics interchangeably. Boeing Aerospace Corporation (BAC) provided a burst processor and actively participated both in test runs on a telescope at Mt. Lemmon and on KAO flights. The burst processor provides a very useful real-time tool as well as a convenient way to store 128 consecutive frames onto an 8" floppy disc. The Boeing signal processing group has done a great deal of analysis of image size and motion as well as having worked on the TDI demonstration. Teledyne Brown Engineering (TBE) contributed two versions of the optical design for the dewar in its f/16 configuration, worked with Aerospace on the f/6.5 optics, provided the germanium lenses, participated in ground and flight tests, did much of the planning of test procedures, and performed major post-flight data analysis. The University of Denver built and flew a cold-optics radiometer to provide a characterization of the atmospheric radiance which was coincident in space and time with the array observations. The Aerospace Corporation served as principal investigator on the experiment, coordinated the various groups, provided cryogenic dewar specifications, installed and aligned the optics in the dewar, conducted the various observing runs and test programs, provided the primary data acquisition system, and converted the various flight tapes into common 9-track tapes. Recently, Aerospace has also begun to play a major role in data analysis and interpretation.

Dewars and Optics

Overall Goals to be Met by the Mechanical and Optical Design

Two major constraints determined the optical/mechanical design of the KITE detector systems. First, the instruments had to be suitable for use on three different telescopes without moving the detector arrays or the internal wiring. These telescopes are: the Kuiper Airborne Observatory or KAO (f/16, 36-inch primary mirror), the Lear Jet Observatory or LJO (f/6.5, 12-inch mirror), and the ground-based Mt. Lemmon NASA telescope (f/16, 60-inch mirror). Second, the infrared background seen by the detector arrays while observing through a telescope must be low enough not to saturate the detector multiplexer. All of

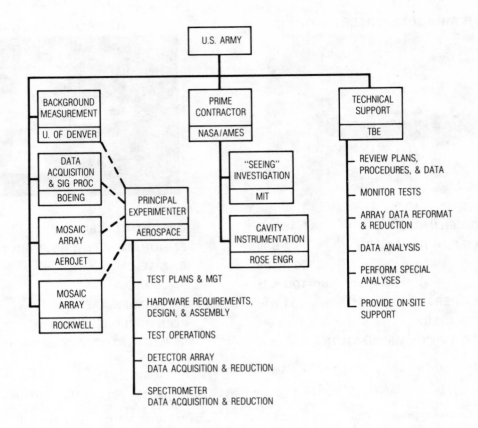

Figure 2.   Organization of the KITE project.

the decisions made during the design and implementation of the KITE systems were ultimately driven by one or the other of these two constraints. In addition we wanted the instruments to be simple to use and maintain and to be portable. For practicality we specified that the cryogenic dewars require re-filling a maximum of once per day.

Cryostats

The detector cryostats had to be small enough to fit into the space available on the airborne platforms (a severe constraint in the case of the Lear Jet), and yet have an interior work space large enough to accommodate the relatively bulky elements required by the optical design. Because time was short, a readily available, reliable, and tested cryostat was strongly preferred. These requirements led us to choose the Infrared Laboratories' model HD-3(8) dewar with an expanded workspace and two vacuum-sealed 41-pin connectors. This liquid-nitrogen shielded dewar has been used by a number of other infrared astronomers who have found it to be reliable. Figure 3 shows a cross-section of the HD-3(8); the liquid-helium-cooled work surface is an 8-inch diameter circle, and there is a vertical space of 4-in between the helium-cooled surface and the nitrogen-cooled radiation shield.

The original TBE optical design called for a 2-inch clear hole through the liquid nitrogen shield which placed an unacceptably high thermal radiation load on the helium cryostat. The current design uses two liquid-nitrogen cooled blocking filters and a cooled rectangular focal plane baffle to reduce the amount of thermal radiation which enters the helium-cooled volume. The helium vessel needs to be refilled only once every 60 hours when the arrays have not been turned on. There is significant electrical energy dissipation from the arrays at the cold work surface when they are operating; in this case the hold time falls to about 24 hours. The thermal radiation shield must be refilled with liquid nitrogen approximately once every 24 hours as well. Thus it is not possible to leave an unused cold dewar unattended for an entire weekend.

The dewar is heavy, about 70 lbs with preamp and mounting hardware, which places severe constraints on the hardware which mates it to the telescope. On an aircraft, the mounting hardware must be capable of maintaining the dewar's optical alignment in the presence of strong vibration. On a ground-based telescope, the dewar's optical alignment with respect to the telescope must be maintained even as the telescope is tilted by up to 60°.

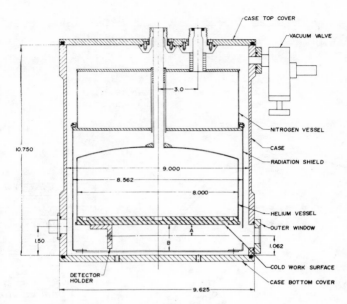

Figure 3. Cross-section of the Infrared Laboratory HD-3(8) dewar used for the detector arrays. Dimensions are in inches. The dimensions marked "A" and "B" are 2" and 4", respectively, for our dewar which necessitated adjustments in some of the other indicated sizes.

Dewar Optical Design

The optics inside the dewar must re-image the focal-plane of the (existing) telescopes onto the detector array at a suitable magnification while introducing as little extra aberration as possible. The design must also include a Lyot stop which is a helium-cooled stop placed at a real image of the pupil to minimize unwanted background radiation reaching the detectors. To minimize thermal background radiation all of the optical train, with the exception of the entrance window, must be cooled to cryogenic temperatures. Additionally, the optics should be kept as simple as possible so that they can be aligned in the laboratory and not lose alignment when cooled or when exposed to shock and vibration on board the aircraft.

The production of compact, high-quality re-imaging optics is a challenging task for the optical designer. Two designs were developed, one for the f/6.5 LJO telescope, the other for use on both the f/16 KAO and the f/16 Mt. Lemmon 60-inch telescope. The f/6.5 design uses a pair of off-axis paraboloids (Figure 4); the system has an overall magnification of unity and serves merely to take the image formed by the telescope focal plane and re-image the light onto the detector focal plane. When used on the LJO 12-inch telescope, each 125 x 125 micron detector subtends an angle of 13 x 13 arcseconds (65 x 65 microradians). The AESC Si:Bi bulk detector array mounted in this optical system was first flown on the LJO in May, 1985 to measure backgrounds and system noise. No star observations were made at that time because of problems with the pointing servo-system of the telescope.

Because we wanted the optical system used on the KAO to have a de-magnification of approximately 2.5, the simple two-mirror system used for the LJO was not adequate. Such a system gives unacceptably large aberrations when used at a magnification other than unity. While Aerospace was awaiting the award of contract, TBE undertook the design of a re-imaging system which would give acceptable optical performance and still fit inside the infrared dewar. Their foresight allowed the acquisition of the optics in a timely fashion and avoided delays in the program. Because of the large amount of special wiring and cabling that is needed to mount the detectors in the dewar, we asked that the new design not require that the detector substrate be moved from its original position on the helium surface. The initial TBE design used a single Ge lens which was mounted just inside the penetration of the liquid nitrogen shield and simply condensed the beam; it is shown in Figure 5. Because there was no Lyot stop, this first design gave a high infrared background which saturated the detector array. The final TBE design is shown in Figure 6. A germanium lens is used to re-image the telescope focal plane and to provide a real image of the pupil for a Lyot stop. Two flat mirrors, one before the lens and one after the lens, fold the optical path so that it fits into the dewar. This optical scheme is used by all three detector systems when they are mounted on the KAO or on the Mt. Lemmon 60-inch telescope. Table 1 summarizes the re-imaging optics and their performance on the telescopes.

Table 1. Performance of Re-Imaging Optics

Telescope	f/ratio	Aperture	Expected Effective Detector Size	Blur Circle[1]
LJO	6.5	12"	~ 70-100 μrad	~ 75-240 μrad
KAO	16	36"	~ 20-24 μrad	~ 40 μrad
Mt. Lemmon	16	60"	~ 13-16 μrad	10-35 μrad

[1]Typical seeing, including diffraction effects, known aberrations, etc.

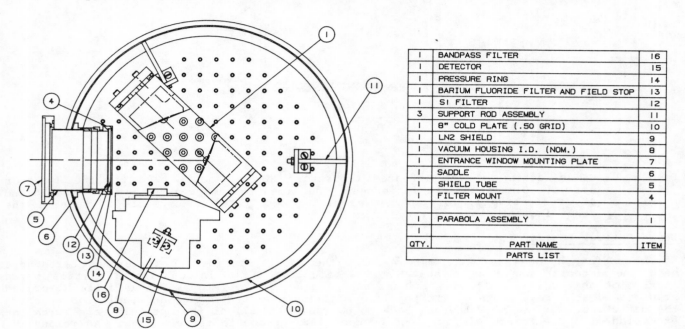

I	BANDPASS FILTER	16
I	DETECTOR	15
I	PRESSURE RING	14
I	BARIUM FLUORIDE FILTER AND FIELD STOP	13
I	SI FILTER	12
3	SUPPORT ROD ASSEMBLY	11
I	8" COLD PLATE (.50 GRID)	10
I	LN2 SHIELD	9
I	VACUUM HOUSING I.D. (NOM.)	8
I	ENTRANCE WINDOW MOUNTING PLATE	7
I	SADDLE	6
I	SHIELD TUBE	5
I	FILTER MOUNT	4
I	PARABOLA ASSEMBLY	I
I		
QTY.	PART NAME	ITEM
	PARTS LIST	

Figure 4. Off-axis paraboloid re-imaging optics used for the LJO f/6.5 telescope. Cold baffles are omitted for clarity. View is from below the dewar (see Figure 3).

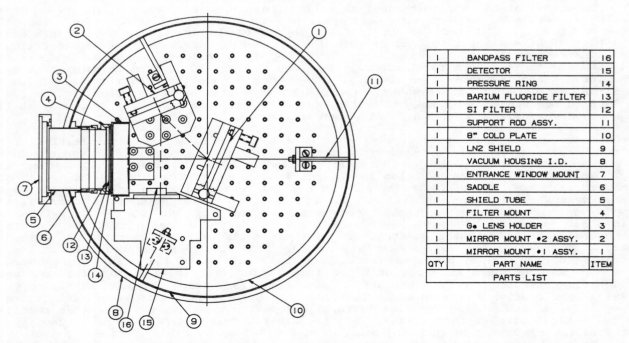

I	BANDPASS FILTER	16
I	DETECTOR	15
I	PRESSURE RING	14
I	BARIUM FLUORIDE FILTER	13
I	SI FILTER	12
I	SUPPORT ROD ASSY.	11
I	8" COLD PLATE	10
I	LN2 SHIELD	9
I	VACUUM HOUSING I.D.	8
I	ENTRANCE WINDOW MOUNT	7
I	SADDLE	6
I	SHIELD TUBE	5
I	FILTER MOUNT	4
I	Ge LENS HOLDER	3
I	MIRROR MOUNT #2 ASSY.	2
I	MIRROR MOUNT #1 ASSY.	I
QTY	PART NAME	ITEM
	PARTS LIST	

Figure 5. Original condensing lens scheme designed but not used at f/16 because it has no Lyot stop and gives a high background. Cold baffles are omitted.

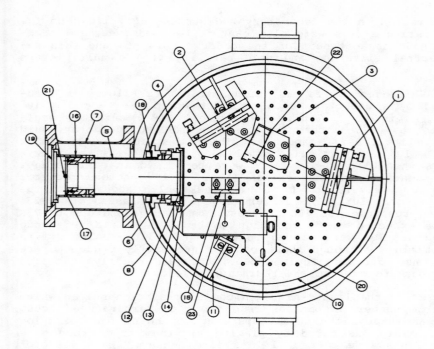

QTY.	PART NAME	ITEM
I	BANDPASS FILTER	23
I	LENS	22
	TELESCOPE FOCUS/DEWAR ROTATION POINT	21
I	ARRAY	20
	ENTRANCE WINDOW MOUNT SURFACE	19
I	GUIDE TUBE KAO/AESC	18
I	APERTURE PLATE KAO/AESC	17
I	SHIELD TUBE SUPPORT KAO/AESC	16
I	PUPIL KAO/AESC	15
I	PRESSURE RING	14
I	BARIUM FLUORIDE FILTER	13
I	SI FILTER	12
3	SUPPORT ROD ASSEMBLY	11
I	8" COLD PLATE (.50 GRID)	10
I	LN2 SHIELD	9
I	VACUUM HOUSING (NOM.)	8
I	EXTENSION ASSY. HD-3(8) DEWAR	7
I	SADDLE KAO/AESC	6
I	SHIELD TUBE KAO/AESC	5
I	FILTER MOUNT	4
I	Ge LENS/MOUNT	3
I	MIRROR MOUNT #2 ASSY.	2
I	MIRROR MOUNT #1 ASSY.	1
QTY.	PART NAME	ITEM
	PARTS LIST	

Figure 6.   Final optics used for the Kuiper Airborne Observatory f/16 36-inch telescope and the Mt. Lemmon Observatory 60-inch telescope. This design re-images the pupil onto a cooled Lyot stop, and greatly reduces the background seen by the array. Helium-temperature baffles are omitted.

The Germanium Lens

A germanium lens has a refractive index high enough to give the short focal length demanded by the final optical design without introducing unacceptable aberrations; it also serves as an additional short-wavelength blocking filter. The use of a germanium lens presents two problems, however. First, because germanium does not transmit visible light, the lens must be aligned by "dead reckoning". Second, an uncoated Ge lens transmits only ~ 40% of the incident radiation. Thus the lens must be anti-reflection (A-R) coated to reduce reflection losses. This A-R coating must adhere to the lens at liquid helium temperatures and survive repeated cycling between 4K and 300K. Thermal tests were made of an A-R coating supplied by CVI Laser Corporation on a germanium flat. Adhesion after three cycles between room temperature and liquid helium temperature was found to be excellent, and the coating readily passed an "adhesive tape" and an "eraser" abrasion test. Unfortunately, the lenses that were subsequently coated shed their A-R coatings after only one thermal cycle. These lenses have since been re-coated with acceptable results, although some of them are starting to shed their A-R coating where they are in contact with the clips that hold them in their optical mounts.

Although the HD-3(8) dewars were specially modified to give extra room in the "vertical" direction (perpendicular to the optical axis), the largest germanium lens that can be fitted into the dewar in a rigid mount without obstructing other parts of the optical path has a diameter of only about 2 inches. This is too small to allow all of the detectors in the Rockwell BIBIB 10 x 50 element array to be illuminated by the entire primary mirror. Even with the optics optimally aligned, detectors at the extreme edges of the 50-element columns are vignetted. Any performance figures that we discuss for this array are for fully illuminated detectors near the optical axis.

Optical Mounts, Baffles, and Background Considerations

Mounts for the optical components were designed and constructed at The Aerospace Corporation. The major issues addressed when designing the mounts were differential thermal contraction and the thermal sinking of elements to the helium surface. Even now the most thermally-distant components cool relatively slowly so that the first 3-liter load of liquid helium is lost within 20 minutes. At the end of this time, however, the optics are sufficiently cold that the second 3-liter load of helium will last almost a full 60 hours.

The requirement for using existing array devices was driven by cost and schedule constraints. For example, it was only about 6 months from the time that the first array was put into a dewar until it had been tested in the laboratory and successfully flown on the LJO and KAO. Unfortunately, the existing arrays were all meant to be used under low background conditions (i.e., had small node capacitances of the order of 0.4 pf) and this required particular attention to spectral filtering and cold baffling if the multiplexers were not to be saturated.

If the background from the liquid-nitrogen cooled radiation shield were allowed to illuminate the entire hemisphere viewed by a detector, each detector would see ~ 1.3×10^8 photons per second per micron at 11 microns. Thus, the Rockwell BIBIB detector would see ~3.6×10^7 photons per second in its 0.28-µm bandpass, and the AESC MC^2 detector would see ~2.5×10^8 photons per second in its 1.9-µm bandpass. The number of photons (7.5 x 10^4) collected in one frame time (3×10^{-4} sec) is quite small when compared to the saturation level of ~ 10^6 electrons. Yet, when blanked-off tests were performed in early stages of the project, the devices were essentially in saturation (i.e., > 10^6 photons/detector per frame). This was traced to a long wavelength (> 18 µm) leak which appeared in the bandpass filters when they were operated at 10K. The filters have a transmission of ~ 25% from 20 to 28 µm, a region of strong response for IBC devices. The background due to 77K radiation through the leak would be expected to be too high to be useful as a reference "zero". Thus, liquid-helium cooled baffles were placed between the helium-cooled optics and the nitrogen-cooled radiation shield. Although these baffles are fairly crude and are fabricated from aluminum shim, aluminum foil and aluminum tape, they are readily installed and removed, and greatly reduce the background from the radiation shield.

We also found several "light leaks" in the liquid-nitrogen cooled radiation shield which were letting in an unacceptably large number of photons from the room-temperature vacuum canister. The largest light leak was through the metal caps over the holes for the nylon stabilizing pins in the baseplate. This was cured by placing aluminum tape around the metal caps. Other small radiation leaks were corrected in a like manner. Each detector currently sees a background flux from the dewar alone of less than 5×10^8 photons per second when the helium-cooled baffles around the optics and the light leaks into the radiation shield are blocked. This is adequate to serve as a zero-incident photon (ZIP) background when compared to the background seen by the arrays through the 10 µm filters.

Windows and Filters

The dewar entrance window is made of ZnSe and has been anti-reflection coated to give maximum transmission near 10 microns. Unlike a number of other commonly used infrared window materials, ZnSe is relatively insensitive to degradation in a humid environment. The material also has a very low infrared emissivity (< 1% at 10 microns), which is particularly important as it must be used at "room" temperature and furthermore fills the beam. One disadvantage of ZnSe is that it has a high refractive index and must be anti-reflection coated to avoid unacceptably high (~ 30%) reflection losses.

Infrared filters are placed at two positions inside the dewar. A set consisting of an 8-13 µm interference filter coupled with an uncoated BaF_2 blocking filter is mounted on the liquid-nitrogen cooled radiation shield (see Figures 4 and 6). This filter set prevents unwanted thermal radiation from entering the liquid-helium cooled areas and protects the detectors from seeing unnecessarily high background radiation; the filters also reduce the thermal radiation load borne by the helium cryostat as discussed above. The BaF_2 filter blocks long-wavelength leaks in the 8-13 µm interference filters and radiates like a 77K blackbody as far as leaks in the narrow filter on the array mount are concerned. A helium-cooled filter (or two in the case of the RI array) is placed directly in front of the detector array. This narrow-band interference filter is the primary determinant of the infrared bandpass seen by the detector. For the Rockwell BIBIB array, the limiting filter has a half-power bandpass from 10.5 to 10.7 µm warm. The AESC arrays use filters with a half-power bandpass from 10.5-12.5 µm warm.

Data Acqusition System

The data acquisition system is based on a 16-bit 8086/8087-based S-100 bus computer system assembled from a combination of boards, some of which were commercially available and some of which were designed and built in-house. The functional diagram is shown in Figure 7. Integrated circuits were specifically chosen for high speed operation (8 MHz) because of the extremely high data rates associated with the experiment. For a 307 µsec frame time and 512 pixels at 12 bits each, the full bandwidth is about 3.3 Mbytes/sec of data. Including housekeeping, the data rate is 5 Mbytes/sec. There is no reasonable way to intercept that data stream with a conventional computer, so a number of interface boards were designed and built at The Aerospace Corporation to send the data, together with housekeeping and time code information, directly to the Ampex 28-track high density recorder while sending in parallel the array data to the host computer. The host computer is capable of acquiring approximately 1 out of every 4 frames sent to it at this rate. The computer can display either a color representation of the data at about 15 Hz, or send the

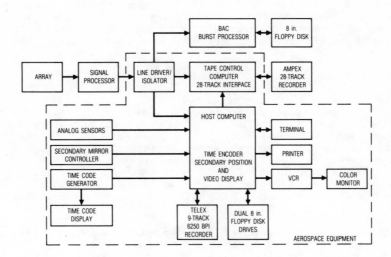

Figure 7. Functional diagram of the 16-bit 8086/8087-based S-100 bus computer system. Aerospace equipment is enclosed by the dotted line; other equipment was GFE or supplied by other contractors.

1 in 4 frames to either a 6250-bpi Telex 9 track tape recorder (3200 frame files) or to an 8" floppy disc (1228 frames/disc). The acquisition of the data in the host computer is done with a buffered interface board and a "frame grabber" (a dual-port memory and video display processor) designed and built at Aerospace. In addition, boards were built to encode the time and the position of the secondary mirror during each frame, to decode the time during 28-track tape playback, to simulate the array output so that hardware and software tests can be done independent of functioning arrays, to interface the 28-track playback electronics to the host computer so that the data quality on the 28-track tapes can be verified, and to interface between the 28-track and 9-track recorders during playback and downloading of the data to 9-track tapes.

Software was written to provide a versatile display in real-time. Corrections for offset and gain variation in the array output (pattern noise) can be applied to the data before display on the color monitor. In addition, the host computer can control the position of the rocking secondary mirror of the telescope providing a background subtraction[1,2] at ~ 27 Hz ("chopping") to enhance a target signal relative to the background. However, on our second KAO series we found a background measurement would typically be valid for up to 10 minutes, which allowed the storage and subtraction of that background without chopping. We believe that the background varied so slowly because of the improvements that NASA had made in the vacuum tightness of the cavity with respect to the plane's cabin; this presented the array with a fairly stable background in the open port cavity. Some residual fluctuations can still be seen in the background data, especially when played back in a "movie" mode on a color terminal. We believe that this is due to small motions of the telescope with respect to the cavity as the servo stabilization system corrects for aircraft motion. That servo correction, incidentally, results in an image motion of < 5 µrad, rms.

To date the most difficult task has been interfacing the 28-track Ampex recorder to the array outputs. A large number of early parts failures made it difficult to get started, but these appear to be resolved now. The recorder requires a pixel clock for control of tape speed, and this clock must meet exacting specifications for symmetry, stability, and absence of noise. Although more than 10^{10} bytes of data were recorded on 28-track tapes during three 7-hour flights, problems with interference on the pixel and frame sync lines have made more than 2/3 of it unrecoverable. Thus far we have used instead the smaller amount of real-time 9-track data which was recorded in parallel. We believe that hardware fixes have corrected this problem for future flights. As it currently exists, line drivers with optical isolators which are capable of driving 3 ports are connected directly to the digital data bus of the array electronics package which was provided by the array manufacturer. One of the outputs is then connected to a buffered interface board which conducts I/O with the time encoder board, the secondary position A/D board, and the frame grabber, which are all located in the host computer. This buffered interface board merges the housekeeping data with the array data and sends it to the 28-track recorder.

We now have two options for playback. We can slow down the 28-track recorder and play the tape directly into the 16-bit system without losing frames, or we can connect the Ampex playback electronics to the buffered interface unit (BIU). The BIU uses a microprocessor as part of a tape controller and resides on two S-100 bus cards. DMA (direct memory access) is used to transfer the data from the 28-track high density tape to a more common 6250 bpi 9-track tape. The BIU uses interrupt-driven software and First-In-First-Out buffers (FIFOs) to increase system throughput. All of these devices have now been run successfully in the laboratory.

Field Tests

When new devices of any sort are created, one of the first tasks undertaken is to attempt to characterize the device under extremely well-controlled conditions. In the case of an IR detector array, this usually means a laboratory environment with the detector installed in a dewar designed specifically for the array in terms of background, spectral filtering, detector temperature, and detector bias. The electronics may be run by batteries, or at a minimum by carefully regulated power supplies, and the entire system is usually in a room where there is no EMI (electro-magnetic interference). This is as it should be, for only in this manner can one map out the performance of the device as a function of the many variables associated with its operation and identify the ultimate limits on its performance.

Unfortunately, this does not address the operating characteristics of devices in the real world. The following plan was therefore devised to move in reasonable increments from the manufacturer's laboratory to airborne observations from the KAO.

The helium dewars were specified several months in advance of the award of contract to Aerospace and delivered on schedule. Next the optics, provided by TBE, were to be installed at Aerospace and the entire dewar taken to the manufacturer for array installation. Plans allowed for several weeks of operation in the Aerospace laboratories following installation. Because of numerous delays the first dewar, containing the AESC bulk array, was delivered the day before departure for the Mt. Lemmon Observatory. At Mt. Lemmon we had to install the dewar in a mounting cradle, set up a test fixture to simulate the telescope, and align the dewar in its cradle. The 60" aperture of the Mt. Lemmon telescope afforded us enough collecting area to see IR stars such as μ Cep or α Ori with a good signal-to-noise ratio. The spot size was essentially as predicted, but the background was unacceptably high. At this point design and procurement of the second version of the f/16 optics were initiated by TBE.

Two off-axis paraboloids, as discussed earlier, were then installed in the dewar and flown on the Learjet Observatory (LJO). The LJO has a very noisy acoustical environment. It also has ground return currents flowing in the skin of the aircraft and frequent radio transmissions by the pilots. Both of these can produce bad electrical interference. We concluded that the LJO flight would be a good test of the impact of an airborne telescope platform on the noise performance of the arrays and their electronics. In flight, the noise spectra of specific pixels were within a factor of two of those measured in the laboratory. Infrared backgrounds were markedly reduced in the air. In retrospect, we believe that the relatively low noise level was probably due to our operating the arrays in a region of soft saturation (slightly too high a bias voltage for the background) where the noise appears artificially small.

Less than one month after the LJO tests, we installed the system on the KAO, which required a complete changeover of the dewar optics and data acquisition systems. The installation of the new optics, which had a liquid-helium-cooled Lyot stop, was prompted by the tests at Mt. Lemmon which had shown an unacceptably high thermal background. During the first two KAO flights, we used the AESC bulk detector array to collect data on response and noise as a function of bias voltage for four IR stars and for Jupiter, which is an extended source. Although we looked for the effects of changing the operating altitude from 39,000 to 45,000 feet, infrared background radiation from the cavity completely dominated all other effects. On the third flight, degradation of the surfaces of the primary and tertiary mirrors increased the background and further degraded the performance of the array.

The RI BIBIB array and AESC MC^2 array were subsequently taken to Mt. Lemmon for shakedown runs and the RI system, including electronics, was flown on the LJO to look for noise problems. Again, only minor problems were noted during these tests, except that we installed a narrower filter (~ .28 μm) for the RI array in order to reduce the background on the detectors. While at Mt. Lemmon, AESC took laboratory array data with the MC^2 detector system viewing liquid nitrogen, dry ice and acetone, and a room-temperature source to provide information for flat-fielding studies. We are just beginning to use these results in our data reduction.

In September-October 1985, a crack was found by NASA in the bulkhead of the KAO telescope cavity. The repair and subsequent testing forced a delay of over 50 days. Following this delay, we conducted a series of three KAO flights in January, 1986 in an attempt to operate all three arrays (two per flight) with the complete data-acquisition system including the 28-track high-density tape recorder, which had not been used on the previous KAO flight series. During the first of these flights the telescope was out of collimation and gave comatic, deformed images. Technical difficulties with the telescope also plagued the second flight, but the third flight was successful. During the seven hours of the third flight, all three dewars were mounted on the telescope in turn and we observed the stars α Ori and CW Leo and the nebula M42, a region with multiple sources and extended emission. During the observations of α Ori with the RI dewar, a scanning experiment sequence was conducted by Boeing and TBE to simulate time-delay-and-integrate (TDI) operation. In this test, the image of the star was scanned across the array at different rates to simulate a moving target. The subsequent analysis of the TDI data by BAC showed an improvement in the signal-to-noise ratio of 2.7 versus the theoretical limit of 3.0 even though the secondary mirror is not designed to do this sort of experiment. To the best of our knowledge, this represents the first demonstration of TDI with an operational system, and is a significant accomplishment of the KITE program.

Preliminary Results

As noted above, we have only begun to scratch the surface of the mass of KITE data, but a few comments may be made at this time. The image core appears to be of the correct size, although the wings of the image extend further than expected. The wings probably do not come from the multiplexer because SWIFET (switched FET) multiplexers are noted for their extremely low crosstalk. We plan further laboratory experiments to assess the effect of misalignment or poor focus of the dewar optics.

While observing a bright celestial point source during the first series of flights we noted a low-level illumination of the focal plane over all but the few rows and columns that were vignetted due to a slight misalignment of the dewar optics. Because the low level flux was visible only on those detectors which were able to see out of the dewar, we concluded that the effect was probably due to scattering off dust on the telescope optics.

To first order, all of the arrays show essentially the same system NEP of ~ 2×10^{-13} W/$\sqrt{Hz}$. But we have noted some non-photon counting sources of noise as we have learned how to optimize the arrays' performance. We are exploring the nature of the noise to look for pick-up (a 120 Hz peak came and went in flight), level shifts such as one might expect from changing out-of-field illumination caused by the motion of the telescope in its cavity, or noise due to ground loops on the airplane. The array processor electronics were constructed with the capability to move the single point ground around the system. However, aircraft schedule constaints have thus far precluded our doing an extensive test of the noise in the aircraft on the runway to study the effects of moving this single ground point or adding additional grounds.

The motion of the image seems minimal, ≤ 2.4 μrad, and to within our ability to measure it may be constant. The spot size and motion data are consistent with the results being obtained at visible wavelengths by Dunham and Elliot[3] as part of a separate seeing study being done under KITE.

We have now written and tested software to transfer data from both the realtime 9-track tapes and the down-loaded 9-track tapes which are generated from the original 28-track tapes. Now that the data-transfer problems are solved, more sophisticated and more efficient data analysis will be done on the laboratory's VAX 11/785 computer. The IDL language which we use for this work has proven to be a very valuable tool due to its flexibility, ease of use, and large built-in capability. The data can be output as files on disc or tape, as color images or as a 3-D representation of the array intensity (Figure 8). Hard copies are easily made. We are also able to compute averages and standard deviations for each pixel, to find the image centroid, and to study noise spectra using FFT's. With Boeing, TBE, Aerospace, and U.S. Army analysts all studying the data, we hope to maximize the value we can derive from the approximately 100 billion bytes obtained thus far.

Acknowledgments

The excellent support of the many people at Ames Research Center and elsewhere who helped with many aspects of the flights is gratefully acknowledged. Particular thanks are due those people at Rockwell International and Aerojet ElectroSystems Company who helped us to get the most out of their arrays. This work was supported by NASA Grant NAS2-12155.

References

1.   Low, F. J. and Rieke, G. H., "The Instrumentation and Techniques of Infrared Photometry," in Methods of Experimental Physics, Vol. 12A: Astrophysics, ed. N. Carleton, Academic Press, 1974.
2.   Allen, D. A., Infrared: The New Astronomy, John Wiley and Sons 1975.
3.   Dunham, E. W. and Elliott, J. L., private communication.

AEROJET ELECTRO SYSTEMS COMPANY BULK ARRAY

CW LEO

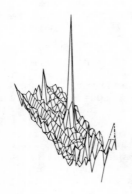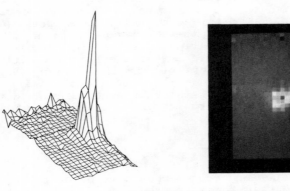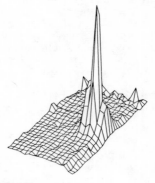

ROCKWELL INTERNATIONAL BIBIB ARRAY

α ORI

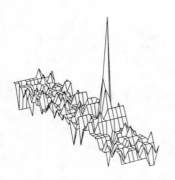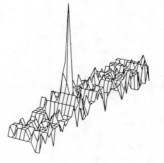

AEROJET ELECTRO SYSTEMS COMPANY MC² ARRAY

α ORI

Figure 8. Typical pixel images of point sources (originals are larger and are produced in color) and 3-D contour representations of output for each of the three arrays tested. The data were obtained during the 29-30 January 1986 flight of the KAO. Note that test conditions (e.g., source, integration time, wavelength interval, etc.) differ greatly for each of the arrays so that comparisons should not be made based on these displays.

INFRARED TECHNOLOGY XII

Volume 685

Session 3

Infrared in the United Kingdom

Chairs
Douglas E. Burgess, Peter N. J. Dennis
Royal Signals and Radar Establishment, United Kingdom

The pyroelectric vidicon; ten years on

D Burgess*, R Nixon[+] and J Ritchie[+]

*Royal Signals and Radar Establishment, Malvern, Worcs, UK

[+]English Electric Valve Company, Chelmsford, Essex, UK

Abstract

Although the pyroelectric vidicon infrared sensitive television camera has been available for over a decade, development of the tubes and cameras did not cease with the demonstration of the first imagery.

This paper describes firstly the progress which has been made in the UK towards a higher performance vidicon tube, and secondly the improvements to cameras which were necessary in order to obtain the optimum performance from the tubes. Operation of the cameras in a number of situations will be described.

Introduction

More than ten years ago pyroelectric vidicon infrared sensitive television camera tubes were fabricated in the UK and other countries. Visible waveband TV cameras were modified to take the special tubes, or built specially, and imaging was performed using simple germanium lenses. Results from these room temperature operating cameras was encouraging, when first soldering irons were seen and later hand silhouettes could be discerned against a cold background[1,2]. All this took place whilst the people concerned had only a limited understanding of how their systems were operating, but it gave them sufficient confidence to proceed to further developments which are described in this paper.

The pyroelectric effect

A temperature dependent, spontaneous polarisation, reducing to zero at a curie temperature, T_c, in a way analagous to magnetism, is known as the pyroelectric effect. Its characteristic is shown in figure 1. Variations in the temperature of a pyroelectric material give rise through the pyroelectric effect to variations of surface charge. By sampling the charge, changes in temperature of the material may be measured. To a first approximation the charge signal is proportional to the rate of change of the material's temperature.

This fundamental property of pyroelectric materials has an important consequence. When used as an infrared detector, a pyroelectric does not respond to stationary thermal scenes, so that a radiation modulator such as a simple chopper is required if static scenes are to be imaged. This disadvantage is counteracted by the advantage that a pyroelectric detector responds only to the AC component of a thermal scene. Unlike a photon detector, a pyroelectric detector array is not required to have the linearity or dynamic range to cope with the very small contrast on a high DC pedestal typical of normal scenes. A further feature of pyroelectric materials is that since they can cool and heat, both negative and positive charges are produced. This property particularly affects the detector readout mechanism if as is the case of a television camera tube, only negatively charged electrons are available to discharge the signals.

Pyroelectric vidicon tube

The construction of a pyroelectric vidicon camera tube is shown in figure 2. It differs mainly in two respects from its visible sensitive counterpart. Firstly a germanium entrance window is fitted to allow transmission of infrared radiation. Secondly a thin slice of a pyroelectric material, triglycine sulphate (TGS) in the early tubes, replaces the photoconductive target behind the window. The TGS is cut and polished with its polarisation axis perpendicular to the face of the slice, enabling pyroelectric charges to be read by the scanning electron beam.

In operation a germanium lens focusses the infrared scene on to the TGS target. As areas of the target heat or cool, the surface charges are neutralised by the electron beam, raster scanned by means of external coils.

TGS was chosen as the first pyroelectric target material for the vidicon tube because of its high figure of merit, p/ε. It has a high pyroelectric coefficient, p, the slope of the polarisation curve of figure 1, and a low dielectric constant, ε[3]. This latter is most important since the impedance of the readout beam is significant and a low capacitance target is required to ensure complete charge readout at the standard 25 Hz television rate. When the readout is not efficient a "laggy" picture results with thermal trails following moving objects[4]. TGS also conveniently has a curie point of 49°C, enabling sensitivity laboratory operation of the tube without the need for a stabilising temperature control.

Since the pyroelectric detector generates both positive and negative charges, a bias mechanism is necessary within the vidicon tube to ensure that the signals can always be discharged by a negative electron beam. A number of bias techniques have been employed, including conducting targets[5] and ramp voltages applied to the signal plate,[6] but those found most suitable have made use of the electron beam during line flyback: in normal television tubes the beam is blanked during flyback to ensure no charge readout.

In the pyroelectric tube two flyback methods have been devised. Firstly in the USA a secondary emission method has been developed, where during the line flyback high velocity electrons impinge on the target with an ejection efficiency greater than unity. The electrons ejected from the target drift towards and land on the positively charged tube grid, leaving the target with a nett positive charge[7]. A different source of positive charge has been developed in the UK[4], France and Germany. Here the vidicon tube is filled with a low pressure gas which is ionised during the line flyback. Those ions generated between the target and the grid drift towards and land on the target, raising its potential. Those generated elsewhere within the tube drift in the opposite direction to bombard the cathode; a potential life-reducing mechanism.

An example of imagery obtained from an early RSRE experimental camera using a pyroelectric vidicon tube built by English Electric Valve Company, UK (EEV) is shown in figure 3. The camera was held stationary whilst the cyclist pedalled through the field of view, creating the time-varying temperatures.

Tube improvements

i. Sensitivity

Imagery such as that of figure 3 gave sufficient confidence to begin a programme of improvements to increase the tube sensitivity. Firstly, alternative pyroelectric materials were considered. In particular it was found that the deuterated form of TGS, DTGS, had not only a lower dielectric constant, and thus a higher p/ε figure of merit than TGS, but also its curie point was higher, expanding the operating temperature range of the tube. Because of its lower capacitance, target thickness could be reduced with DTGS to give a higher sensitivity without a loss of readout efficiency. Deuterated triglyciene fluoroberyllate has also been considered as a target material since at first sight it has an even higher figure of merit than DTGS. It does however suffer from toxicity problems and a more rigorous analysis of performance which derives a different figure of merit shows that the sensitivity gain is only marginal.

ii. Resolution

The limiting 200 lines resolution of the DTGS tubes was caused by sideways heat spread in the pyroelectric target. In order to increase resolution it was necessary to dice the target into small islands separated by a low conductivity medium. A reticulation programme was set in motion aimed at ion beam machining a pattern of narrow grooves at 30 to 40 microns pitch all over the target. Before machining, the target complete with signal plate had to be stuck to a backing layer to ensure its continuity after reticulation. Figure 4 shows a cross section of a reticulated target. Grooves a few microns wide are machined completely through to the backing layer.

Measurements have been made to compare the performance of tubes with reticulated and standard targets in a camera using a radiation chopper which allowed stationary patterns to be observed[9]. Figure 5 shows the modulation transfer function results for the two tubes. Despite these results being obtained several years apart, they can still be compared to show the considerable improvement in MTF brought about by the reticulation process. At first it was feared that reticulated targets would have a reduced low spatial frequency response because of the loss of absorbing material from the grooves. In practice this effect has proved to be negligible.

Camera improvements

i. Choppers

For some thermal imaging tasks involving only a qualitative analysis, a simple pyroelectric camera can be used in the panned mode. For more quantitative work however, particularly when the more sensitive DTGS tubes became available, the problems of image smear associated with the panning mode of operation led to a decision to incorporate a rotating disc chopper in a new experimental camera. Photographs of the simple panned camera and the experimental camera containing a chopper are shown in figure 6.

Cameras which are fitted with choppers are more suited to precise measurements of temperature[4]. Static scenes can be imaged and if necessary temporal integration can be used to further enhance a camera's temperature sensitivity. Along with the introduction of a chopper comes increased electronics complexity since alternate open and closed chopper periods (cameras are normally run with one TV field scan with the chopper open, the next with it closed) give rise to alternate polarity signals for each image pixel. Signal inversion on alternate fields is necessary as shown schematically in figure 7.

ii. Flicker processing

The effect of alternate field signal inversion is not only to give a constant polarity signal for real temperature differences but to give alternating polarity outputs for any tube artifacts such as large area shading and small area blemishes. The resulting displayed image exhibits an annoying 25 Hz flicker which can confuse detail in a scene of low temperature differentials. In order to minimise this problem a post-camera electronic processor (a flicker processor) has been developed which incorporates a one field delay and a differencing circuit (figure 8) to remove all video components which are of opposite polarity on successive fields[10]. Not only does this unit remove the effects of spatial shading, it also reduces the trailing thermal effects from hot moving objects. Originally this electronics occupied a considerable volume; with improved memory chips it can now be housed inside a camera.

iii. Readout uniformity

With the introduction of a chopper, signal inverter and a flicker processor, the pyroelectric camera may be used in earnest to measure temperature differentials and to determine temperature differences between image pixels. Only by the use of these items to improve camera performance was it possible to investigate thoroughly the topic of signal uniformity across the pyroelectric tube.

Pyroelectric targets are made to a high degree of parallelism, lenses are designed to have negligible vignetting, but still imagery such as that of figure 9 was obtained under some circumstances. The reduced brightness around the periphery is accompanied by a loss of signal in the same area. The imagery can be explained by a less effective beam readout towards the target edges resulting in a reduced signal firstly from variations of temperature within the scene and secondly from any temperature differential between the warm chopper enclosed in the camera and the cooler scene. On opening the camera case the chopper attains the ambient temperature and the brightness variation disappears, although the signal reduction is still present.

A study of the relative positions of the tube, the scanning coils and the focus coil was carried out to determine if an optimum existed which would maximise the beam landing dynamics over the whole target. Figure 10 shows plots of signal across a target diameter for a range of coil positions. A sharp optimum is evident which has proved reproducible in production.

RSRE camera developments

Following on from the cameras shown in figure 6, RSRE undertook a programme to develop a smaller, lower power camera with automatic operation to enable it to be operated by unskilled personnel. Figure 11 is a photograph of the latest camera which incorporates all the improvements described above. In order to ensure ease of operation in the field, the power consumption has been kept below six watts without compromising the performance. For comparison, imagery of the same bicycle and rider as that shown in figure 3 is shown in figure 12, using a reticulated DTGS tube made eight years ago.

Microphony has been a problem with some pyroelectric detectors. The reticulated tubes were found to be much less sensitive to vibration than were the standard tubes. We were able to hard mount the camera to a helicopter platform without problems and obtained microphony-free images.

Industrial camera development

Much of the development of pyroelectric cameras in the UK during the last few years has taken place in Industry, most notably at EEV and at Insight Vision Systems. EEV became particularly interested in the pyroelectric camera for firefighting after successful trials of an early experimental model. Their development has now led to a large scale production capability of environmentally-protected cameras, complete with viewfinders, which have been bought in very large numbers by both civilian and military firefighting teams (figure 13). Large sums of money have been saved in the UK by the ability of firemen to navigate through smoke filled buildings to find the seat of a fire. At least one life has been saved by one of these cameras being used to detect the heat pattern of a woman trapped under fallen masonry after a gas explosion had caused an apartment block to collapse.

EEV have also developed a range of cameras for laboratory and scientific purposes covering wavelengths from visible out to hundreds of microns. Since the pyroelectric detector can be sensitive to all wavelengths, it is even possible to image different wavelengths simultaneously.

Insight Vision Systems have concentrated their development work on cameras suitable for scientific and industrial applications. Facilities such as temperature contouring have been included to ease image analysis, whilst a boresited visible camera has been shown to be useful in aiding the location of hot spots. Figure 14 shows an Insight camera system complete with viewfinder and comprehensive controls. This equipment is now vying with systems from AGA for the market in light weight portable thermal cameras with image analysis facilities.

Conclusions

The first pyroelectric vidicon camera pictures were, by today's standards, poor in both thermal sensitivity and spatial resolution. However the results were sufficiently exciting to convince the experimenters that further development was desirable. In a programme that has lasted more than ten years that early promise has been transformed by a successful development programme into systems of real scientific, industrial and safety worth. Evidence of the usefulness and importance of the devices may be seen from the sales figures of the companies now involved in their manufacturing. The development, particularly of complete systems, has not yet come to an end. We shall await the next ten years with interest.

Acknowledgments

The support of RSRE and of the English Electric Valve Company Limited for the pursuance of this work is gratefully acknowledged.

References

1.   Infrared pick-up tube with electronic scanning and uncooled target. D R Charles, F Le Carvennec. Advances in electronics and electron physics Vol 33A, 1972, p 279-284.
2.   Thermal imaging with pyroelectric television tubes. E H Putley, R Watton. Advances in electronics and electron physics Vol 33A, 1972, p 285-292.
3.   Pyroelectric materials: operation and performance in thermal imaging camera tubes and detector arrays. R Watton. Ferroelectrics Vol 10, 1976.
4.   The pyroelectric vidicon: a new technique in thermography and thermal imaging. R Watton, D Burgess, B Harper. Journal of applied science and engineering A, Vol 2, 1977, p 47-63.
5.   Pyroelectric materials for operation in a hard vacuum pyroelectric vidicon. R Watton, G R Jones, C Smith. Advances in electronics and electron physics, Vol 40A, 1976, A301-312.
6.   Ramp mode operation of a pyroelectric vidicon. A L Harmer. IEEE Transactions on electron devices, Vol ED-23, 1976, No 12.
7.   Theory and performance of pyroelectric vidicon with an electronically generated pedestal current for the cathode potential stabilised mode. T Conklin, B Singer, M H Crowell, R Kurczewski. IEEE International electron devices meeting, Washington, 1974, p 451-454.
8.   Infrared television: thermal imaging with the pyroelectric vidicon. R Watton. Physics Technology, Vol 11, 1980, p 62-66.
9.   The thermal behaviour of reticulated targets in the pyroelectric vidicon. R Watton, D Burgess, P Nelson. Infrared physics, Vol 19, 1979, p 683-688.
10.   Improved performance from pyroelectric vidicons by image-difference processing. C N Helmick, W H Woodworth. Ferroelectrics Vol 11, No 1-2, p 309-13.

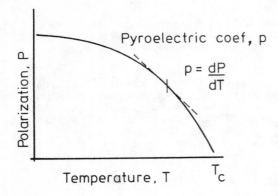

$$p = \frac{dP}{dT}$$

Pyroelectric coef, p

Figure 1. Pyroelectric effect

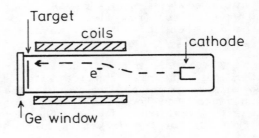

Figure 2. Pyroelectric tube schematic

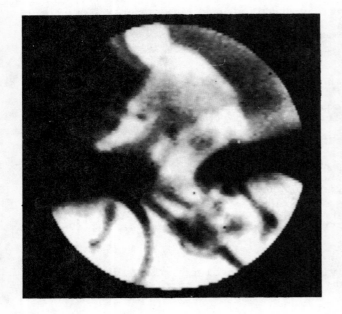

Figure 3. Early imagery of a cyclist

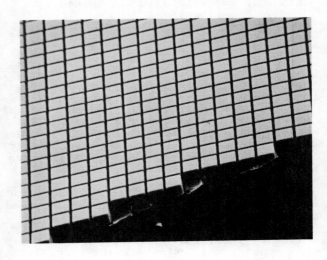

Figure 4. Reticulated pyroelectric
vidicon target

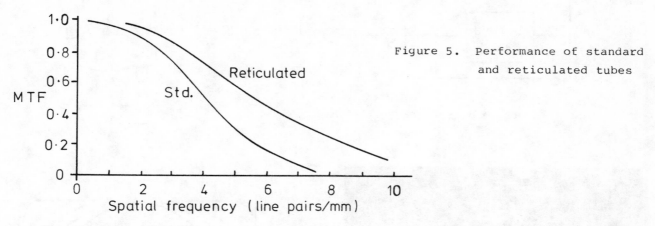

Figure 5. Performance of standard and reticulated tubes

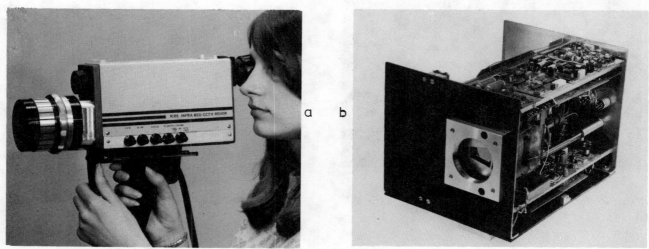

Figure 6. RSRE pyroelectric cameras: (a) early simple design (b) first camera with chopper

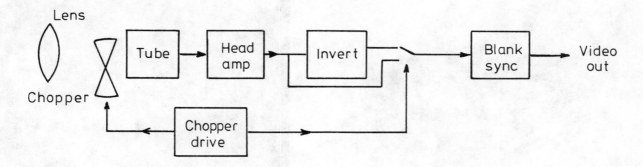

Figure 7. Signal path in a camera using a chopper

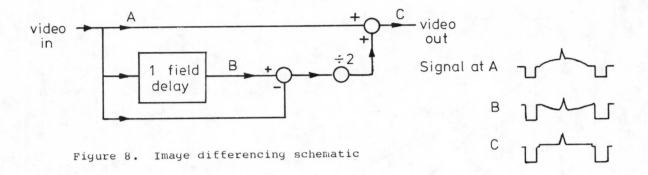

Figure 8. Image differencing schematic

Signal at A

B

C

Figure 9. Example of non-uniform signal readout

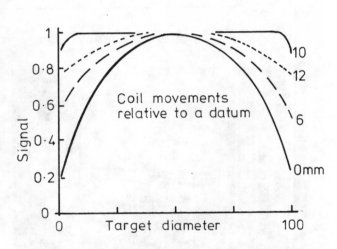

Figure 10. Optimisation of signal readout uniformity

Figure 11. Most recent RSRE camera

Figure 12. Cyclist imaged now

Figure 14.   Insight vision systems camera

Figure 13.   EEV firefighting camera

An Uncooled Linescan Thermal Imager for Ground and Airborne Use

T J Liddicoat, M V Mansi
Plessey Research Roke Manor Limited
Roke Manor, Romsey, Hampshire   SO51 OZN U.K.

D E Burgess, P A Manning
Royal Signals and Radar Establishment
St Andrews Road, Malvern, Worcestershire WR14 3PS U.K.

Abstract

The Plessey Company, in conjunction with the U.K. Ministry of Defence, has developed a range of pyroelectric infrared detectors which in linear and two dimensional array formats may be used at room temperature. Sensor research based upon these devices is carried out at the Royal Signals and Radar Establishment (RSRE) Malvern, and Plessey (Roke Manor). This paper describes sensors employing linear arrays of detectors, and gives examples of imagery that has been obtained.

The sensors built to date employ germanium lenses, a chopper to modulate the incoming radiation, and sixty four element linear arrays of detectors. The Plessey system reads the peak value of signal from the detector with the chopper "open" and displays this in the form of a grey-scale picture, whilst the RSRE system takes the difference between the signal with the chopper open and closed and displays this as a grey-scale picture.

For the systems discussed, the instantaneous field of view is about 3.5mR and the NETD (Noise Equivalent Temperature Difference) better than 0.2K.

The sensors may be used in a panned mode to give large-area surveillance, in a staring mode to provide image information by virtue of the motion of the target or mounted in an aircraft in a linescan mode. Examples of imagery obtained in these modes are presented.

Introduction

High performance thermal imaging systems are now available with displayed picture quality equal to, or in some cases exceeding that normally associated with good quality monochrome television systems. Such thermal imagers (e.g. U.K. Common Module, U.S. Common Module) are expensive both in terms of initial cost and logistical in-service support. The high initial cost is contributed to by the cost of cryogenically cooled detectors, whilst the in-service costs result from the need to keep the detector supplied with cryogen, whether it be in the form of air stored at high pressure in rechargeable bottles, local compressors or cooling engines.

Whilst there is clearly an on-going need for high performance thermal imaging systems, there is an increasing requirement for low cost sensors which do not necessarily have to represent the ultimate in sensitivity and spatial resolution. Such sensors may be used in remotely piloted vehicles for airborne surveillance, in security systems and in remote ground sensing applications. The sensors may be used to present an image to a human observer, but there are many applications where the output is required to be processed automatically.

In the search for a low cost thermal imaging sensor, the removal of the requirement to cool the detector is the first major step. Plessey Research are sponsored by the U.K. MoD to develop arrays of pyroelectric detectors, which operate at local ambient temperature in the 8-14 micrometre waveband. The Company also currently has a privately funded research programme which is investigating the use of pyroelectric detectors in thermal imaging sensors. A U.K. MoD programme of systems research is also being carried out by the Royal Signals and Radar Establishment.

As part of these U.K. research programmes, thermal imaging sensor technology demonstrators have been built employing linear arrays of pyroelectric sensors. A sensor built by RSRE was described in a paper presented at the SPIE Cannes conference (Reference 1) in November 1985. Another sensor, built by Plessey, was briefly described in a paper presented at an IEE conference in June 1986 (Reference 2).

Sensor Design

Pyroelectric detectors

Characteristics   The pyroelectric effect has been observed for many years, and is discussed in relation to practical detectors in, for example, Reference 3.   The critical aspect of the detector from the system designers' point of view is that pyroelectric detectors, in common with other thermal detectors, are essentially a.c. coupled devices; that is they respond to changes in the incident radiation.   The implication of this is that the scene must be scanned by the detector, or the radiation incident on the detector has to be modulated by some form of chopper, be it mechanical or electro-optic.

Linear arrays of detectors   The Plessey Company have developed linear arrays of detectors, up to 64 elements are currently available, in which each element is compensated by an element which is shielded from incident radiation, and is buffered by a JFET source follower mounted in the detector package.   Adjacent elements are isolated by a saw slot to reduce thermal crosstalk.   The detector array is mounted on an anti-microphony mount (Reference 4), and individual detector lead-outs are brought to a standard connector on the mounting assembly (Figure 1).

The individual detector element D* is typically greater than $10^8 cmHz^{0.5}W^{-1}$ in the 8-14 micrometre waveband, operating over the frequency range of 10 to 200Hz.

Overall system design

The overall concept for both sensors is that the radiation collected by a germanium lens is modulated at 50Hz by a mechanical chopper mounted just in front of the image plane (i.e. the detector).   The output from the detector is amplified, multiplexed, converted to a digital format, processed, scan converted and displayed on a TV monitor for visual assessment.   Both sensors use the same type of detector.   There are, however, fundamental differences in the way the information from the detector is treated in the two systems.

The RSRE sensor   In the RSRE sensor, the output from each detector element is A.C. coupled to a buffer amplifier, multiplexed onto a single line, filtered, sampled for chopper open and closed fields, digitised, image difference processed, scan converted and displayed (Reference 1).

The Plessey sensor   In this sensor, the outputs from each detector element is coupled to a bandpass filter centred on the chopping frequency, amplified, multiplexed onto one line, amplified, sampled on peak of open field, digitised, scan converted and displayed.

Detail description of the Plessey sensor
The system block diagram is shown in Figure 2.

Optical head   Infrared radiation from the scene is focused on to the detector array by a germanium lens and modulated at 50 hertz by a two bladed chopper mounted immediately in front of the detector package.   The array and chopper are configured so that the detector outputs are temporally in phase, and the chopper speed is locked to a video reference so that the detector outputs and the final video display output are synchronised.   The lens used is a 50mm focal length, f/0.8.   The detector elements are 180 m by 140 m separated by a 40 m slot.

Preamplifiers   The preamplifiers provide noise band limiting and gain.   Although direct multiplexing of the detector outputs is feasible, difficulties arise in maintaining detector limited noise performance and accommodating the differential dc offsets from the JFETs in the particular detector package being used.   The addition of gain and ac coupling between the detectors and multiplexer significantly relaxes the multiplexer specification in terms of noise performance and dynamic range.   The preamplifiers consist of single operational amplifiers with a bandpass characteristic centred on 50 hertz.

Multiplexing and Digitization   The 64 preamplifier outputs are multiplexed into a serial stream of analogue data which is digitized by a fast analogue to digital converter.   The multiplexer switches the output at a fast rate so that all the detector outputs are sampled in a short time interval centred on the peak of the modulated signal waveform.   In this way, all the detectors are sampled at a point at which their output is at least 97% of their peak output.

Scan conversion   The digitised array samples are stored in a solid state memory, thus building up a two dimensional image from successive linear cross sections of the thermal scene.   The imaging head is normally oriented so that the array views a vertical section in the scene.   The captured image data is converted into a standard video output to provide a real time display.   Data can also be transferred to computer disk for storage or direct to a processor.

<u>Sensor performance</u>   The imaging system has been designed to ensure that the final image quality is limited only by detector noise.   The calculated NETD, assuming this condition, is less than 0.15K.   This figure is the result of equating the response from the predicted optical power incident on the detectors with the total integrated detector noise (including noise band limiting by the preamplifier).   Measurements of the system front end performance have shown that the predicted NETD has been achieved and system MRTD measurements, using human observers viewing the video display, have also been carried out.

The modulation transfer function due to spatial sampling of the scene is given by the product of two sinc functions, one describing the spatial sampling due to the finite detector size, and the other the effect of spatial separation of successive samples.   In the vertical direction (along the array axis) the latter quantity is fixed by the detector pitch and is therefore constant for a given array configuration.   In the horizontal (scan) direction, however, the separation of successive samples is determined by the scan rate, the temporal sampling rate being constant at 50 hertz.   The consequent dependence of MRTD on the scan rate is illustrated in Figure 3 which shows the result of the MRTD measurements using 50mm focal length optics and a vertical bar pattern test target. Curves for two scan rates (6.7 and 14.4 degrees per second) are shown; for contiguous samples, the scan rate is approximately 10 degrees per second.   It can be seen that at the lower target spatial frequencies, the MRTD for both scan rates is similar but, as expected, the knee in the MRTD curve for the faster scan rate is at a lower spatial frequency.

An example of the imagery produced by the experimental sensor is shown in Figures 4, 5, 6 and 7.   The prototype sensor head is shown in Figure 8.   The banding, particularly evident on high contrast edges, is due to absorption of some incident radiation by the compensating elements in the detector package.   This problem has been resolved by more effective masking in subsequent devices.   It should be noted that the image shown is a direct representation of the detector outputs subject only to conditioning by the preamplifiers.   No processing or correction for non-uniformity of responsivity is necessary to yield an image of acceptable quality and the preamplifiers employ fixed value components and are not matched to individual detectors.

Discussion

Relative merits of the RSRE/Plessey sensors

Both the sensors demonstrate similar performance, as may be judged from the picture of a Landrover taken with the Plessey sensor (Figure 7) compared with that taken with the RSRE sensor (Figure 9).   However, the theoretical (and measured) NETD of the Plessey sensor is slightly less than that for the RSRE sensor, primarily because the noise bandwidth of the Plessey design is less than that of the RSRE design.   The principal advantage of the RSRE design is a reduction of the front end component count, which becomes important for sensors employing long linear arrays of detectors.   However, it is a matter for the system designer to trade the relative merits of the two approaches according to the particular requirement.   If small signal multiplexing is to be used, an image difference processor (IDP) must be used to remove the offsets inherent in the multiplexer.   If the multiplexer can be used later in the system, where the offset variations are small compared with the detector signals, IDP need not be used.

Alternative approach

Systems which do not use choppers have been considered.   The function of the chopper is two-fold.

Firstly, it shifts the signal modulation frequency to a region where it may be conveniently handled.

Secondly, it provides an absolute radiometric reference for the detector, and therefore provides an extra dimension of data to any following signal processor.

In an unchopped system design, it is necessary to consider carefully the requirements of the sensor.   Since the detector is AC coupled, it will generate a differential signal as it is scanned from one scene artefact to the next.   If a reasonable degree of radiometric referencing of adjacent pixels is required, it is likely that in most systems the detector will be required to operate down to very low temporal frequencies, which may pose significant problems to the electronic design.

Applications

This type of sensor is ideally suited to slow scan area monitoring, low resolution aerial reconaissance, route monitoring and slow scan radiometric measurement in, for example, medical thermography or industrial plant thermography.

Figure 4 is an example of an image taken by passing the sensor across a scene of interest. This class of use would be applicable to security applications with an automatic processor to detector intruders.

Figure 5 is an example of a thermographic image of a man's face. The quality of the reproduction does not do justice to the information present in the original. (The figure is a grey scale rendition of a colour photograph). However, the inherent low cost and simplicity of the uncooled sensor could mean that a useful tool could be available to every medical practitioner.

Figure 6 is a thermograph of a building. The construction industry should find such a low cost, simple sensor useful for thermal surveying purposes.

Figure 7 shows an image obtained by moving the object past a stationary sensor. Clearly such images may be presented to an automatic processor which could carry out classification functions. Automatic traffic monitoring by roadside sensors is therefore an economic possibility.

Figure 9 is another example of a panned image, taken with the RSRE sensor.

Figure 10 shows an image obtained from the RSRE sensor mounted in an aircraft. With the limited resolution available (64 lines), there is still useful data present. The prospect of longer linear arrays will allow the design of useful low cost sensors which may be used in remotely piloted vehicles.

Framing Sensors

Currently the upper useful temporal frequency limit for the detectors is about 300Hz, although operation at higher frequencies is possible if reduced thermal sensitivity is acceptable.

For contiguous spatial sampling with a sensor chopped at 300Hz the pixel sampling rate is 300 pixels/second.

For an unchopped sensor the equivalent pixel rate for a contiguous spatial sampling is, for an upper temporal frequency limit of 300Hz, 600 picture points per second.

It is therefore possible to conceive framing sensors which use current generation linear array detectors which operate at up to 600 scanned pixels per second. There are many slow scan T.V. applications which can use this type of sensor.

Future Developments

Two important developments are required for linear arrays of detectors.

The first is to increase the number of elements in the package. This will enable the design of simple, high spatial resolution sensors.

The second is to reduce the size of the detector elements whilst maintaining sensitivity. This will enable the optical system to be scaled in size accordingly, with a consequent saving in weight and cost of the optics.

Further developments will incorporate multiplexers into the array package, to enable lower cost systems' electronics to be used.

Conclusion

Thermal imaging sensors have been designed, built and demonstrated which use totally uncooled pyroelectric thermal detectors in linear array formats.

The sensors employ different signal conditioning and processing electronics which demonstrate two approaches to a system implementation, either of which may be used according to the specific requirements and constraints placed upon the system designer.

Both sensors are radiometric, in that they may be calibrated to give absolute information concerning the thermal signature of the scene.

Both sensors are the basis for miniaturisation and should offer a low cost, medium performance thermal imaging capability.

Acknowledgements

The authors wish to acknowledge the co-operation between the U.K. MoD and Plessey Research and Technology Limited, for support in carrying out the detector and sensor research reported here.

References

1. Manning, P. et al, SPIE Infrared Technology and Applications, 590, 2, (1985).
2. Liddicoat, T.J. et al, IEE Conference Publication No. 263, P34, (1986).
3. Porter, S.G., Ferroelectrics, 33, 193 (1981).
4. Shorrocks, N.M., SPIE Recent Developments in Materials and Detectors for the Infrared, 588 (1985).

Figure 1   Detector Assembly

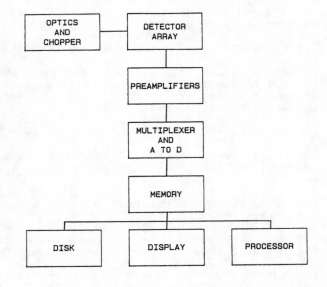

Figure 2   Block Diagram of the Plessey Sensor

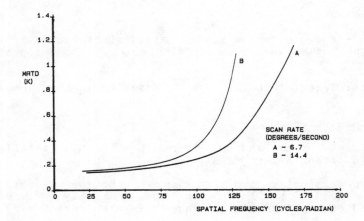

Figure 3   Minimum Resolvable Temperature Difference (MRTD) vs. spatial frequency for the Plessey Sensor

Figure 4   Panned Image (Plessey Sensor)

Figure 5
Panned Thermogram of a Man's Face (Plessey Sensor)

Figure 6
Panned Thermogram of a Building (Plessey Sensor)

Figure 7   Image produced by target movement

Figure 8   The Plessey Prototype Sensor Head

Figure 10   Pushbroom Linescan Image (RSRE Sensor)

Figure 9   Panned Image (RSRE Sensor)

Two dimensional infrared focal plane arrays utilising a direct inject input scheme

R. A. Ballingall and I. D. Blenkinsop

Royal Signals and Radar Establishment, Malvern, Worcestershire, UK

I. Baker and J. Parsons

Mullard Limited, Southampton, UK

Abstract

This paper describes two different focal plane arrays, both using direct injection inputs. We compare a device using a charge coupled device multiplexer with a more novel multiplexed electronically scanned array (MESA), which can give better performance in the 8-10 μm band. The basic limitations of the direct injection technique are discussed, including sensitivity to gate bias voltage and to I·R product of the infrared diode. I·R product is proposed as a figure of merit. Charge storage capacity and dynamic range are also discussed, especially in relation to the performance required of the uniformity correction electronics.

Introduction

In this paper we describe two dimensional CMT-silicon hybrid focal plane detector arrays utilising a direct inject input scheme. This offers a compact, low power solution. Two different readout schemes are presented, each of which may be operated in either the 3-5 μm or 8-10 μm waveband at 77K. We compare a conventional CCD readout scheme with a more novel multiplexed electronically scanned array (MESA). The basic limitations to the direct injection mechanism are considered and current-resistance product is proposed as a figure of merit.

We derive theoretical performance figures for the arrays described and consider the stability levels required of the direct inject gate voltage. Finally we briefly consider the dynamic range of the output signals and the required performance of the uniformity correction electronics.

The use of current-resistance product as a figure of merit

The transfer of photocharge from an infrared diode into a silicon readout circuit can be characterised by the low frequency injection efficiency η. In order to minimise the noise of the transfer circuit, a high injection efficiency is needed. The low frequency injection efficiency can be expressed in terms of the diode resistance, R, and the transconductance, g_m, of the direct injection device.[1,2]

$$\eta = \frac{g_m R}{1 + g_m R}$$

The resistance of the infrared diode will vary with the amount of bias applied to the diode. For a diffusion limited diode

$$I = I_O \left(\exp \frac{qV}{kT} - 1 \right) - I_p$$

$$R = R_O \exp - \frac{qV}{kT}$$

$$I_L = \frac{kT}{qR_O} \left(\exp \frac{qV}{kT} - 1 \right) \quad .$$

The transconductance of the silicon interface circuit is a function of the photocurrent I_p and the leakage current of the diode I_L

$$g_m = \frac{q(I_L - I_p)}{nkT}$$

$$g_m R = \frac{1}{n} \left[1 - \left(\frac{qI_p R_O}{kT} + 1 \right) \exp - \frac{qV}{kT} \right] \tag{1}$$

To maximise the injection efficiency the value of $g_m \cdot R$ should be as high as possible and this requires a high $I \cdot R$ product. Alternatively the above conditions can be expressed entirely in terms of the diode properties

$$IR = \frac{kT}{q} \left[1 - \left(\frac{qI_p R_o}{kT} + 1 \right) \exp - \frac{qV}{kT} \right] \tag{2}$$

The application of reverse bias to the infrared detector will increase the $g_m \cdot R$ product ($I \cdot R$) and hence the injection efficiency. Figure 1 shows the increase in $I \cdot R$ product with reverse bias, V_R, for various values of $I_p \cdot R_o$ product. The graph shows a continuing increase in $I \cdot R$ product with reverse bias. In practice this will not happen because the infrared diodes depart from the diffusion limit at some value of reverse bias that is dependent on the cut-off wavelength. At a cut-off wavelength of 10 μm the peak resistance occurs at a few tens of millivolts. Figure 2 shows a typical result. The other factor limiting the operating point is the increase in 1/f noise with reverse bias. In practice the onset of 1/f noise corresponds closely with the bias level at which the diodes depart from the diffusion limit.[3]

The use of $I_p \cdot R_o$ as a figure of merit is not limited to systems which use the direct inject concept. In the more general case the normalised signal to noise ratio (D_λ^*) is determined by the noise terms and by $R_{I\lambda}$ the current responsivity

$$D_\lambda^* = \frac{R_{I\lambda} A^{\frac{1}{2}}}{\left[\frac{2kT}{R_o} \left(\exp \frac{qV}{kT} + 1 \right) + 2qI_p \right]^{\frac{1}{2}}} \tag{3}$$

To demonstrate the usefulness of $I_p R_o$ as a figure of merit, consider the zero bias case. If the two noise current sources are made equal in (3) then

$$I \cdot R = \frac{kT}{q} (1 + e^{qV/kT})$$

ie for zero bias

$$I_p \cdot R_o = \frac{2kT}{q}$$

Similarly in (1), for an injection efficiency of 50% at zero bias

$$g_m R = \frac{1}{n} \cdot \frac{q}{kT} \cdot I_p \cdot R_o = 1$$

$$\therefore \quad I_p \cdot R_o = \frac{nkT}{q}$$

In either of these cases the threshold of acceptable performance is given by the simple condition

$$I_p \cdot R_o \geqslant \frac{2kT}{q} \tag{4}$$

From the point of view of diode quality and the signal to noise ratio of a direct injection system, the $I_p \cdot R_o$ product represents a good figure of merit.

One of the main limitations of a direct injection system is determined by the transistor that transfers charge from the infrared diode to the storage capacitor (potential well in the CCD). This transistor operates in common gate configuration (direct injection input) and, as such, its transconductance (input conductance in this case) is determined by the current from the IR diode. For a diffusion limited diode, this current depends on the value of I_p (see (2)), which is itself determined by the photon flux from the infrared scene. Figure 3 shows the behaviour of $I \cdot R$ product with cut-off wavelength for diodes at zero bias and with a F/no = 1.4. The upper line is for the best currently available diodes. The lower line is a realistic estimate of the performance of large, two dimensional arrays, where allowances must be made for diodes at the lower end of the distribution of $I \cdot R$ values. This gives a realistic limit for the wavelength of operation of ≃ 10 μm.

The application of reverse bias increases the $I \cdot R$ value of the detectors. For high quality diodes (upper line in Figure 3) this increase can be a factor of 20, allowing operation out to 12.5 μm. For lower values of $I_p \cdot R_o$ the potential improvement gained by reverse biassing is reduced.

This limitation does not apply to diodes operating in the 3-5 μm waveband at 77K provided that the aperture of the optical system is not very small. It is easy to achieve $I_p \cdot R_0$ products of greater than 1 volt and, with the application of reverse bias, the $I \cdot R$ product can be increased further.

Direct injection imaging devices

We have shown that direct injection is possible in both the wavelength bands. The most obvious choice of silicon readout circuit is a CCD. However, close packed two dimensional CCDs suffer the problem of limited charge storage capacity and this can lead to unacceptably short integration times in the long wavelength band. An alternative to this type of array is the MESA.

Multiplexed electronically scanned array - MESA

MESA is a development of the electronically line sampled array concept.[4,5] The MESA consists of a two dimensional array of infrared photovoltaic diodes electronically addressed by MOS switches. Each line of the array is switched in turn to a line of common gate MOS transistors (direct inject), which transfers the photocharge to a line of on-chip storage capacitors (see Figure 4). The voltage on these capacitors is then multiplexed to a single output.

In operation the capacitors are charged to a fixed reset voltage, V_s (~ 5 V). The photocurrent from a line of infrared diodes is then permitted to discharge the capacitors for a fixed time, t, through the direct inject transistors.

The multiplexer is used to transfer the remaining voltages on the capacitors to the output and to reset each in turn to V_s before repeating the process for the next line. The voltage, V_O, appearing at the output is given by

$$V_O = \frac{C_s V_s - It}{C_O + C_s}$$

where C_O is the multiplexer node capacitance which must be smaller than C_s. Using silicon processing with a geometry of 2 μm we have made MESA devices with up to 128 x 128 elements. The constraint on node capacitance, C_O, has been resolved by the adoption of reduced fabrication geometry and a more sophisticated multiplexer design.

A refinement of the MESA scheme is shown in Figure 5. In this case two sets of direct inject transistors and storage capacitors are used. This permits one set to be read out while the other set is accumulating charge from the photodiodes. In this way the output data rate may be reduced while achieving full line equivalent performance. Figure 6 shows a current, typical device with a cut-off wavelength of 10.1 μm.

Charge coupled device - CCD

This type of device has the same constraints on the input structure as the MESA device. It behaves as an MOS transistor in common gate with a CCD well acting as a drain. There is an additional limitation to the close packed, two dimensional CCD. The area of silicon available for the storage element is limited to about one third of the area of the infrared detector. In order to maximise the storage capacity we have used a surface channel peristaltic structure for the close packed section of the device with a high speed buried channel structure for the output multiplexer.[6] This gives a storage capacity of 5×10^6 electrons per detector (48 μm pitch) and this permits a maximum integration time of 8 μs for a 10 μm cut-off detector viewing a 300K scene with F/1.4 optics. The present output data rate is limited to 10 MHz by a combination of the CCD structure and the dewar. This results in a readout time of 400 μs for a 64 x 64 and 1.6 ms for a 128 x 128 array. Figure 7 shows a 64 x 64 array of this type.

The structures used do not permit simultaneous integration and readout. This means that in the long wavelength band, for example, using a 64 x 64 array, integration is possible for only 8 μs out of a total frame time of 408 μs, resulting in approximately one line equivalent performance. As the array size is increased the readout time also increases and thus performance falls below line equivalent. To achieve the above performance, the readout time must be as short as possible and so a number of frames must be summed in an external digital frame store to provide the required output frame rate.

In the 3-5 μm waveband the photon flux is approximately two orders of magnitude lower for similar conditions. This permits the CCD to approach fully staring performance. In the example given above, approximately 2/3 fully staring performance could be achieved.

Comparison of MESA and CCD

The MESA uses much simpler fabrication technology than the CCD. It requires lower driving voltages and fewer, lower speed clock signals. MESA is more tolerant to individual low quality diodes in the array. Such diodes with a high leakage current can lead to saturation of an entire line in a CCD, but only of a single pixel in MESA. Because only a simple switch is needed beneath each detector, the detector pitch may be reduced to the diffraction limit of the optics. This allows more elements to be fabricated from a given size piece of CMT material.

The principal limitation of the MESA structure is that it achieves only one line equivalent performance. However in the long wavelength band this may be better than the performance of a CCD.

The main advantage of the two dimensional CCD is that it can give close to fully staring performance, but only in the 3-5 μm band. In addition the CCD can potentially offer a very high data rate and this may be of value in applications where a "snapshot" facility is desirable.

These considerations make MESA more attractive for the long wavelength band and CCD preferable for the medium wavelengths.

Stability considerations of the direct inject gate voltage

The function of the direct injection transistor is to present a low impedance to the infrared diode and to transfer current to the storage node with minimum noise. The success of this function depends critically on the gate bias potential. The temperature sensitivity of the array can be limited by the stability of this gate bias voltage. From Figure 8

$$dV_g = (R_s + 1/g_m) \, dI$$

where: $dI = I \cdot C_p \cdot \Delta T$

I = total current flowing to the source

ΔT = temperature resolution

C_p = photon contrast.

Figures 9 and 10 plot this for the two wavebands, where cut-on filters of 8 μm and 3 μm respectively are used.

For detectors with high injection efficiency $R_s > 1/g_m$

$$dV_g \leqslant (I \cdot R_s) \, C_p \cdot \Delta T$$

The required gate stability in both wavebands is given for typical cases in Table 1.

Table 1.

	3-5 μm	8-10 μm
I·R product (V)	1.0	0.1
C_p (% K^{-1})	3.75	1.8
ΔT (K)	0.005	0.01
dV_g (μV)	190	200

Care must be taken to ensure that this level of stability is achieved at the focal plane or else the performance will be reduced. If the direct injection transistor is also required to act as a switch, the problems are exacerbated since the bias voltage must also be pulsed. This problem can be eased by interposing a separate switching transistor (or gate electrode for a CCD), thereby separating the injection and switching functions. This is shown (dotted) in Figure 8 and has been incorporated in these chips.

Theoretical performance limitations

It is instructive to consider the theoretical limits to performance of both the fully staring and electronically scanned arrays. In both cases the performance depends on the integration time, t_i. The measure of performance used for an imaging array is the noise equivalent temperature difference, NETD

$$NETD = \frac{4F^2 \cdot B^{\frac{1}{2}}}{M* \cdot A_O^{\frac{1}{2}}}$$

where: F   is the f number of the cold shield
       B   is the bandwisth
       A_O is the optical area of the detector
       M*   is the radiation function (see reference 7)

Figure 11 plots the NETD of an ideal, background limited array versus integration time for f/1.4 in both wavebands. It can be seen that the theoretical values of NETD for both a fully staring (F/S) 3-5 μm CCD and a line equivalent 8-10 μm MESA are between 1 and 3 mK. It must be emphasised that the values plotted in Figure 11 are theoretical numbers and in practice at least a factor of x5 should be allowed to account for such things as imperfect optics, the noise factor of the electronics, a diode detectivity less than the background limit etc. If the theoretical NETD values are ever to be approached, the values of ΔT given in Table 1 would need to be improved, hence imposing even more severe requirements on the stability of the gate bias. Even allowing for these limiting factors the sensitivity of the arrays is still extremely good, being less than 0.02K.

Uniformity correction

The outputs from the focal plane arrays discussed here are not sufficiently uniform to be used without some electronic correction before display. The non-uniformities arise from both the CMT elements and from the silicon readout circuitry. These contribute both fixed offsets and variations in responsivity.

The basic dynamic range required for a background limited system is set by the statistical variations on the number of electrons accumulated at the storage node. For an 8-10 μm MESA this results in a dynamic range equivalent to 14 bits. Close packed CCDs, having a smaller charge storage capacity, require only 11 bits. If the dynamic range within the infrared scene is small enough it may be possible to subtract a fixed DC pedestal from the signal. If the pedestal is large enough, this may allow fewer bits to be used in the digital correction electronics.

In those cases where frame summing is used to match the focal plane data rate to the display rate, greater resolution will be required in the electronics following the A/D converter.

Conclusion

We have demonstrated the value of $I_p \cdot R_O$ product as a figure of merit for direct inject readout schemes. Under some conditions the stability of the direct inject gate bias is the limiting factor on performance.

Using the diode technology available to us, direct injection readout operation is possible in either of the two wavebands considered. We conclude that the CCD is most suitable for the 3-5 μm band and MESA for the 8-10 μm band. In either case a thermal sensitivity of better than 20 milli-Kelvin should be achieved.

Acknowledgements

The authors wish to acknowledge the useful contributions made by M. D. Jenner and G. Crimes of Mullard Limited in the preparation of this paper.

References

1. Sato Iwasa; Director coupling of five micrometer (Hg,Cd)Te Photovoltaic detector and a CCD multiplexer; Optical Engineering, Vol. 16, No. 3, pp. 233-6.
2. Yi Xin Jian; The injection efficiency of direct injection, Infrared charge coupled devices, Infrared Physics, Vol. 21, pp. 53-4.
3. Tobin, Iwasa and Tredwell; 1/f noise in (Hg,Cd)Te Photodiodes; IEEE Trans. on Electron. Devices, Vol. ED-27, No. 1, pp. 43-8.
4. Ballingall, R., Infrared hybrid CMT photovoltaic electronically scanned arrays, IEE Conf. on advanced infrared detectors and systems 1981, Pub. No. 204.
5. Ballingall, Blenkinsop, Lees, Baker, Jenner and Locket; Two dimensional, random access infrared arrays, IEE Conf. on advanced infrared detectors and systems 1983, Pub. No. 228.
6. Baker, Jenner, Parsons, Ballingall, Blenkinsop and Firkins; Photovoltaic CMT-CCD Hybrids, IEE Conf. on advanced infrared detectors and systems 1983, Pub. No. 228.
7. Longshore, Raimondi and Lumpkin; Selection of detector peak wavelength for optimum infrared system performance, Infrared Physics, 1976, Vol. 16, pp. 639-47.

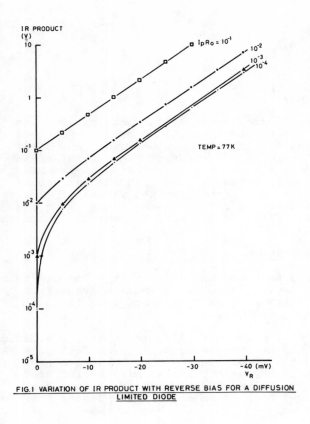

FIG.1 VARIATION OF IR PRODUCT WITH REVERSE BIAS FOR A DIFFUSION LIMITED DIODE

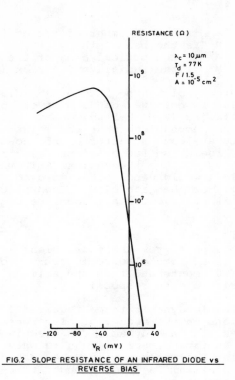

FIG.2 SLOPE RESISTANCE OF AN INFRARED DIODE vs REVERSE BIAS

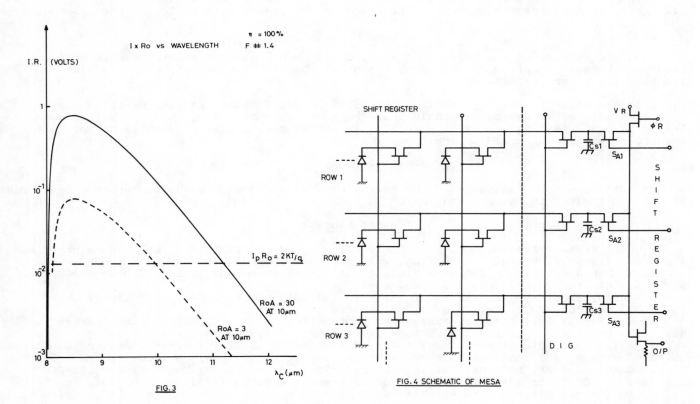

FIG. 3

FIG. 4 SCHEMATIC OF MESA

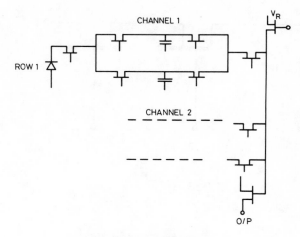

FIG. 5 SCHEMATIC OF MK II MESA

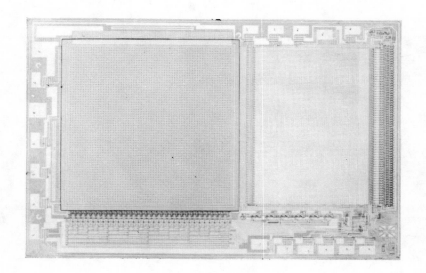

FIG.6 64x64 MESA
10.1μm CUT—OFF WAVELENGTH

FIG.7 64x64 CCD
4.2μm CUT—OFF WAVELENGTH

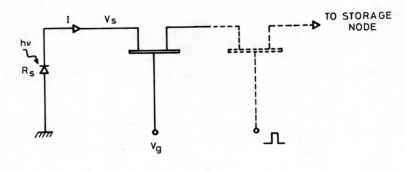

FIG. 8

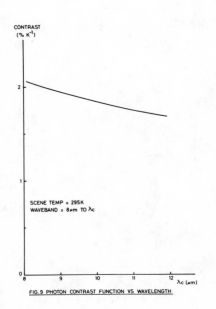

FIG.9 PHOTON CONTRAST FUNCTION VS WAVELENGTH

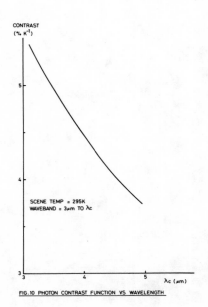

FIG.10 PHOTON CONTRAST FUNCTION VS WAVELENGTH

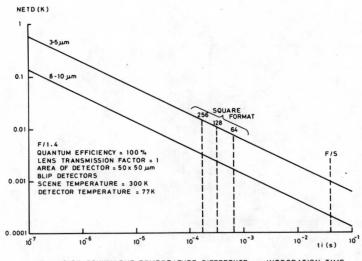

FIG.11 NOISE EQUIVALENT TEMPERATURE DIFFERENCE vs INTEGRATION TIME

Sampling effects in CdHgTe focal plane arrays - practical results

R. J. Dann, S. R. Carpenter, C. Seamer

Marconi Command and Control Systems Limited, Frimley, Camberley, Surrey, U.K, GU16 5PE

P. N. J. Dennis, D. J. Bradley

Royal Signals and Radar Establishment, Malvern, Worcs, U.K

Abstract

An earlier paper by Bradley and Dennis[1] outlined how sampling effects limit the high frequency spatial performance of close-packed two-dimensional focal plane arrays, and described conceptually how these might be overcome. These concepts have since been put into practice and results are presented which show:

a) manifestations of the basic effects.

b) their reduction through microscanning in a real-time imaging environment.

The results were obtained from an imaging system developed by MCCS for RSRE, capable of driving the complete range of 2D detectors currently produced in the UK.

Introduction

The development of close-packed two-dimensional arrays of infrared detectors now offers the potential for simplified thermal imaging equipment[2]. It is therefore possible to envisage a system with no mechanical scanning, as shown in Figure 1.

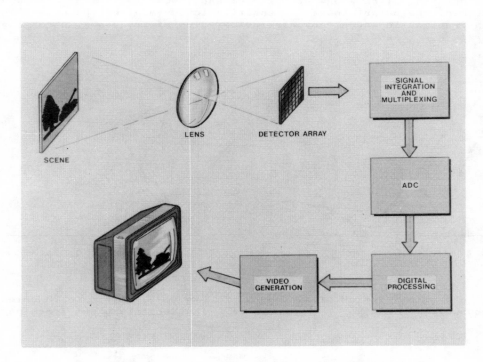

Figure 1. Staring array system schematic

The radiation is imaged through a simple lens onto a two dimensional detector array such that each picture element in the scene is focussed onto a corresponding detector element. Thus, for a given spatial resolution, the field of view will be determined by the number of elements in the array. The simplified opto-mechanics of this system should produce a cheaper, lightweight and more reliable imager than was previously possible.

However, one of the most significant benefits of a staring array is the high thermal sensitivity that may be achieved by virtue of its high stare efficiency. In most scanned systems the scene dwell time on each pixel is much less than 1% of the frame period, but for a two-dimensional array of several thousand elements, this may be increased to 50% or more with a corresponding increase in sensitivity.

A staring array system has been designed and built by MCCS Limited for RSRE in the UK. It uses cooled Cadmium Mercury Telluride (CMT) matrix arrays operating in both the 3-5um and 8-12um wavebands. The detector arrays have been jointly developed by RSRE and Mullard Limited and at present operate at 77K. A CCD readout technology is used in the short wavelength band and co-ordinate addressing for the 8-12um region. The system has produced excellent quality imaging in both wavebands using arrays of up to 64 x 64 elements.

Sampling Effects

Thermal imaging systems, which incorporate unscanned two-dimensional detector arrays at the focal plane, exhibit several effects which are caused by the sampling of the scene by the array[3,4]. High frequency scene detail is obscured in the image by aliasing, causing this information to appear at lower spatial frequencies with a confusing result.

Consider a sinusoidally varying intensity distribution of frequency f which is sampled by a one-dimensional array of detectors whose responsivity profiles are rectangular. If the detector array is correlated with the intensity distribution, the signal from each detector can be determined and the Modulation Transfer Function (MTF) is given by the ratio of image contrast to object contrast such that

$$MTF = \frac{S1 - S2}{S1 + S2}$$

where S1 and S2 are the maximum and minimum values of the detector output as a function of position.

This may be plotted as a function of the frequency and a typical MTF curve is shown in Figure 2 for a detector pitch and width of 40um and a 50mm focal length lens with unity MTF. Because the form of the output image is sensitive to movement of the object the system is said to be non-isoplanatic and at any particular value of frequency, a range of modulation can occur between MTF_{max} and MTF_{min}.

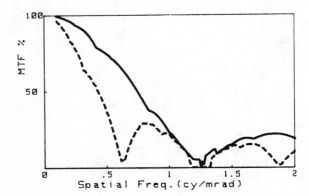 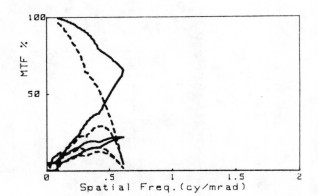

Figure 2. MTF as a function of sinusoidal input frequency.

Figure 3. MTF as a function of fundamental output frequency.

——————— MTF_{max} ------- MTF_{min}

If the MTF is plotted as a function of the fundamental image frequency as shown in Figure 3, it is apparent that the highest spatial frequency which can be reproduced is equal to half the sampling frequency. This is called the Nyquist frequency, N_f, and all frequencies above this are aliased back into the 0 to N_f region, losing all scene information above this frequency limit.

Microscan

In order to reduce these effects the possibility of introducing a dither or microscan movement into the system has been investigated[1]. This causes the image scene to be displaced by some fraction of a pixel with respect to the detector array, such that inter-pixel sampling occurs in both horizontal and vertical directions, as shown in Figure 4. This shows nine elements of an array to which 2 x 2 microscan is applied; field 1 records the image at some reference position on the focal plane; field 2 records the image when displaced by half a pixel to the right; field 3 is displaced half a pixel vertically from the previous position and field 4 half a pixel to the left. The microscan system is then returned to its original position ready for the next frame. In a similar manner, a 3 x 3 microscan can be implemented whereby the image is displaced by one third of a pixel for each step.

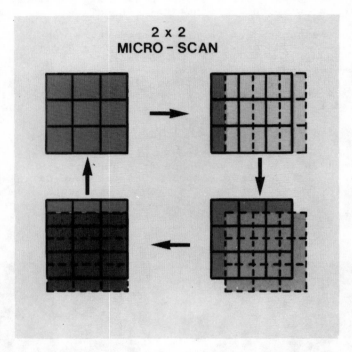

Figure 4.   2 x 2 microscan pattern

Using the formula defined above to calculate the MTF of the system, the improvements obtained using a 2 x 2 microscan are shown in Figure 5 (plotted for one dimension only). It can be seen that the Nyquist frequency has increased from 0.625 to 1.25 cycles/mrad and the level of aliasing reduced significantly. In addition, the system is more isoplanatic, as seen by the reduced spread in MTF.

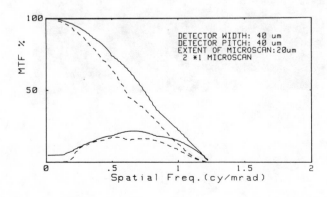

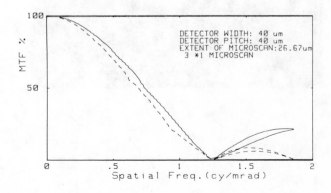

Figure 5.   MTF with 2 x 1 microscan   Figure 6.   MTF with 3 x 1 microscan

————— MTF$_{max}$ --------MTF$_{min}$

Corresponding increases are obtained if a 3 x 3 pattern is generated, giving a Nyquist frequency of 1.875 cycles/mrad and a further reduction in aliasing of approximately two (Figure 6).

These benefits obtained in increased spatial and MTF performance have been achieved at the expense of thermal sensitivity. If a constant frame rate is maintained, the dwell time will be reduced by a factor of four and nine for 2 x 2 and 3 x 3 microscan respectively, hence reducing thermal sensitivity by factors of two and three. However, as previously described, these systems offer the potential for very high thermal resolution and hence this may be considered to be an acceptable trade-off for many applications.

Practical Implementation

The in-built flexibility of the array system previously mentioned has enabled the addition of several microscan modes for practical evaluation.

A versatile microscan mechanism has been developed which uses a single plane mirror actuated by piezo-ceramic transducers. A short focal length detector lens is now acceptable for many applications through the enhanced resolution offered by microscan. This has the added advantage that the mirror, and hence the mechanism, may be of small size with a corresponding rapid response time. System timing is such that a 25Hz display frame rate is maintained for both 2 x 2 and 3 x 3 microscan modes. Video control logic and its associated digital frame store readily adapts to the resulting 128 x 128 and 192 x 192 image formats.

Results

Photographs of bar patterns imaged in normal (non-microscan) and 2 x 2 microscan modes are reproduced in Figures 7a and 7b. Corresponding bar patterns in normal and 3 x 3 microscan modes are shown in Figures 8a and 8b. The improvement in spatial frequency response is clear; the increased isoplanatism of microscanned images is more easily observed on moving targets where the relative phase of the pattern and the detector elements are changing. Measured MTFs show the expected increase in spatial frequency response and correlate well with the theoretical predictions.

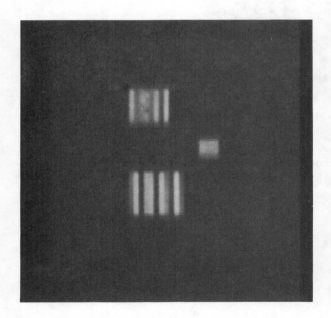

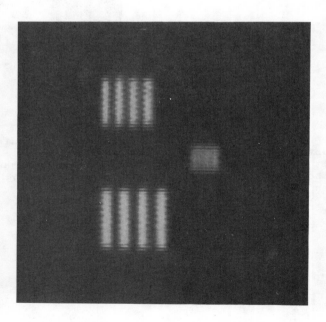

a)   b)

Figure 7. Bar patterns imaged in normal and 2 x 2 microscan modes.

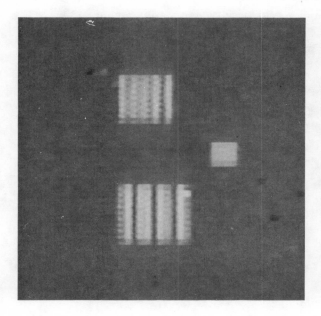
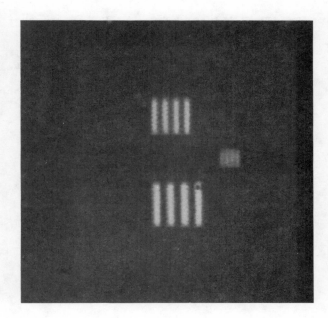

a)   b)

Figure 8.   Bar patterns imaged in normal and 3 x 3 microscan modes.

Conclusions

A practical implementation of microscanning has been demonstrated which validates the theory put forward by Bradley and Dennis[1].   Key benefits are:

* smaller, lighter optics for a given spatial resolution

* reduced aliasing effects

* improved isoplanatism

* wider range of applications (more pixels)

Additional scan modes have also been implemented which further enhance the benefits of the microscan techniques described whilst maintaining the advantages of a simple concept with few moving parts.

Further study is continuing to better define the performance trade-offs under a wider variety of operational scenarios.

Acknowledgements

The authors gratefully acknowledge the support of RSRE and of Marconi Command and Control Systems Limited for the persuance of this work.

References

1. Bradley, D.J. and Dennis, P.N.J., <u>Proceedings of SPIE</u> Vol 590, 1985, Cannes.

2. Dann, R.J. and Dennis, P.N.J., <u>Proceedings of SPIE</u> Vol 572, 1985, San Diego.

3. Duane Montgomery, W., <u>J Opt Soc Amer</u>, Vol 65, 6, pp700, 1975.

4. Witterstein, W., Fontanella, J.C., Newbery, A.R. and Baars, J., <u>Optica-Acta</u> Vol 29, 1, pp41-50, 1982.

A dual waveband imaging radiometer

S.P. Braim and G.M. Cuthbertson*

Royal Signals and Radar Establishment. Malvern. UK
and *GEC Avionics Ltd. Basildon. UK

Abstract

This paper describes a Dual Waveband Imaging Radiometer (DUWIR) which GEC Avionics have developed for RSRE Malvern and which has now been put into production.

This calibrated imager basically provides synchronous, simultaneous imagery in both infrared wavebands, that is pixel by pixel overlay, from a single optical channel. It combines current generation modular FLIR performance with the necessary internal calibrated references to enable it to function as a radiometer, measuring apparent temperatures in both bands. It also has all the necessary radiometric and system status information displayed in graphic form, overlaid on video to permit ease of subsequent analysis of the imagery.

It is based on the Class II, UK Thermal Imaging Common Module (TICM) modules (Figure 1), which comprise the six processing cards and power supply module, the scanner module, the detector lens module, the head amplifier module and the motor drive module which form the basis of the UK high performance, indirect view, modular thermal imager. Selected systems contractors can configure a complete imaging system from these modules by adding a suitable housing, telescope, display, controls and cryogen supply. Note that as the Class II system is intrinsically TV compatible, scanning one IR line for every TV line (including interlaced fields) it does not require any additional scan convertor or interpolator.

Introduction

Essentially the DUWIR(Figure 2)comprises a Class II indirect view opto-mechanical scanner assembly derived from the UK Thermal Imaging Common Module (TICM) Programme[1,2], which has been modified to operate in both wavebands, combined with two signal processing chains from the same class of system (Figure 3). To these are added synchronising and overlay correction circuitry, a comprehensive graphics capability and the necessary acquisition circuitry, drive stages and active devices to proivde the required thermal reference functions in the system. With a suitable control unit this combination produces TV compatible, simultaneous, synchronous imagery in both the 3-5 and 8-12 micron bands, where all the required radiometric and systems status information is overlaid on the video. This permits subsequent analysis of the apparent temperature at any and every point in the image.

Being based on the UK TICM Class II imager, which uses a Sprite detector array, the equipment has the added advantage of inheriting the cosmetically uniform image and near perfect radiometry for which the Class II system is so well known and also the automatic gain (temperature window) and offset (black level temperature) "hands off operation" feature pioneered by the Class II system. The equipment can be operated in wide angle 60° x 40° (including graphics) or in different fields of view by suitable choice of afocal telescope.

This paper describes the philosophy behind the Dual Waveband Imaging Radiometer, the hardware and some representative results obtained using the system.

Philosophy

The requirement for an imaging radiometer stemmed from a growing need for target signature information in both wavebands and a recognition that the demands of instrumentation and subsequent analysis of data from trials was far from trivial. The hundreds if not thousands of next to useless uncalibrated video tapes held across the world containing valuable trials information whose data is in effect irreducible bear witness to this.

In 1981 RSRE defined a requirement for a research tool to measure target apparent temperatures, to assess the relative merits of the two wavebands under conditions where definitive conclusions might be drawn and to assess the benefits of combining the two bands.

The aim was to form a database which would assist in optimising current generation systems and, perhaps more importantly, to assist in determining the direction of future imaging and EO systems.

Whilst there are some excellent industrial/medical radiometers which have been widely used in the past, none of them provide, simultaneously in both wavebands, the required combination of a performance equal to current generation FLIRs, the simplicity of recording and record keeping and the ease of subsequent analysis. Special hardware was therefore required.

In developing the theme, thoughts initially turned to a special to type development. Such a development would have been costly and would, without additional development and environmental test programmes, have risked producing an equipment more fit for the lab than for field trials. Further, it would not have been cost effective to miniaturise the system by the use of hybridisation, which would have further militated against its deployment in a trials environment.

An alternative route was therefore necessary. It was decided to start with an imager which had already been put through development and into production, and to modify it to produce an imaging radiometer. The logical choice was to modify a UK Class II TICM imager. This unit was TV compatible, was in production and had already demonstrated its ability to survive the military environment. It was almost completely hybridised and was therefore small, and its performance was truly representative of the best currently achievable. Further, and most importantly in terms of the conversion, the intrinsic imaging radiometry is near perfect. There are no vignettes, the pupil does not scan across the stop and, during clamping (referencing), scene radiance is completely and exclusively replaced by radiation from the clamp (reference) surface. Modifications of such a unit to provide calibrated imagery was therefore relatively straightforward. There was a high probability of achieving the required calibration accuracy and the build standard of the hardware would inevitably be more highly developed than could be achieved in a special to type development.

Modification of a Class II system was therefore the route proposed.

After a competitive tendering process, it became the selected solution. The brief was to produce the best uniformity and radiometry without substantial changes to TICM Module suite. As will be demonstrated, that objective has been handsomely achieved.

The Hardware

As produced, the equipment comprises a scanner head, a processing electronic unit (PEU) and a dual monitor console incorporating the system control panel and a video mixing facility (Figure 4). A separate keyboard is provided for the insertion of textual comments onto the video channels. (A small semi-ruggedised control panel is available as an option). Figure 6 shows the system in detail.

The scanner head compromises the scanner module itself, the associated motor drive module, the detector lens, the detector, head amplifiers, Peltier cells and their drive stages and cryogenics (in this case a Joule-Thompson mini-cooler), all contained in a suitable environmental enclosure. The scanner module is a standard TICM Class II, modified to include two Peltier references and associated temperature sensors. These Peltiers cells are placed at the intermediate image plane in the scanner and are viewed at the start and the end of the field scan respectively. They can therefore be displayed in the active video period. The standard 8-12 micron refracting optics are replaced with specially designed dual waveband (3-5, 8-12) optics which themselves are coated with customised dual band coatings. Further, subtle changes were made to the reflecting optics to eliminate the possibility of non-uniformity during the clamp period, of a level insignificant when used as an imager but which could have been the limiting factor on uniformity and/or calibration accuracy when used as a radiometer. A wide angle front window is provided. To avoid any possibility of narcissus, the window is tilted such that the detector never sees a reflection of itself. When an afocal telescope is used, the window may be dispensed with. Here we see one of the advantages of the system. As only one telescope is required for both bands there are distinct advantages in terms of size, cost and weight in real installations compared with other dual waveband, dual optic systems. The motor drive package is a standard TICM module. The standard 8-12 micron 8 element Sprite detector has been replaced by a new array which has two, side by side, 8 element Sprite arrays on a common focal plane (Figure 7). One is a standard 8-12 micron TICM array, the other is specially fabricated from 77K 3-5 micron material. Being in a single dewar only one cooler is required, especially advantageous where cooling engines are used and also registration of the images is guaranteed by the geometry of the array. Both arrays use TICM head amplifiers, one of which has had its bias range extended to encompass the range expected of the 3-5 micron array. Note it is here the second subtlety occurs. The TICM imager scan format is one of a family of mechanisms that scan great circle/small circle projections. The off axis array scans a slightly different projection and this has been corrected in the electronics to achieve perfect pixel registration in the two output channels. Less subtly, the delay between the signals from the two arrays also has been eliminated, again in the electronics.

Finally, there is a Peltier cell driver and sense stage. Peltier cells in the scanner head are driven from automatically or manually derived demand waveforms, and are controlled by a microprocessor in a closed loop mode. Platinum resistors on the active surface provide the required feedback. Generally the Peltier cells are driven to near TV waveform black level black on one channel and near white on the other. Other modes of operation also exist. By careful choice of device and calibration good absolute accuracy is achieved and the relative accuracy is exceptional.

The PEU comprises two standard TICM environmental housings, cast side by side (Figure 4). Each side contains a virtually standard TICM signal processing chain (Figure 3) that is an automatic gain matching module, a gain control and clamp module, two bandstore modules (which in paged fashion, convert the eight parallel IR signals to a serial TV compatible signal), a TV processor module (modified), a timing module and a power supply module. Inter-spaced between these modules are the additional modules (Figure 6). Firstly there is a second timing module which synchronises the two channels, overlays the signals by delaying one signal channel in the appropriate bandstores, and also controls the overlay correction by fine timing changes. Secondly there is a graphics module, which generates alpha numeric signals to enable system parametric information and user generated comments to be displayed (Figure 5). Finally, there is an auto control module which provides manual and automatic control of the Peltier cells and acts as a controller for the control panel data bus. For convenience, it has also taken over the automatic gain and offset functions of both channels which are normally embodied in the respective video processor modules. By combining automatic Peltier control with the existing automatic gain and offset, complete "hands-off" operation is available.

To the user, the most obvious manifestation of the changes to the signal processing channel is the graphics display (Figure 5). All the relevant system parameters and status flags are displayed in the video period and the user can insert comments from the keyboard. Additionally, and of more functional importance, direct views of the two primary Peltier references are available in the top left and lower right of the display. Secondary references are injected down the vertical edges of the field to ease subsequent analysis. Along the top and bottom of the image there are grey scales.

In operation, therefore, one can simply interpolate between the two line period secondary references to determine any apparent temperature in the scene. If absolute accuracy is required, one must start by looking at the primary references, then check any discrepancies between those and the secondary references, and check finally any errors in transfer characteristic introduced by, for instance, the video recorder, (by analysising the grey scales). The combination of these steps permits interpolation, to a high degree of accuracy, to the apparent temperature of any point in the scene.

The system control console (Figure 4 and Figure 6) provides two TV monitors, a conventional TV mixer panel with various "Mix", "Wipe" and "Key" functions, and all the controls necessary to operate both IR channels and both Peltier references. A built-in microcontroller communicates with the microprocessor in the PEU. A small, ruggedised, system control panel is available as an option. Indeed both control panels can be connected simultaneously with control priority being determined by the last "local request" pressed on the control panel.

Analysis hardware

Since the output from DUWIR is in UK-standard television format (625 line, 50Hz), a commercial television frame store can be used to capture and hold selected thermal images for analysis. An Intellect 100 frame-store, interfaced to a micro PDP-11 computer, is therefore used for picture storage and retrieval and simple image analysis. A block diagram of the analysis system and its associated peripherals is shown in Figure 8. The 10MByte removable disc unit is used as the prime storage medium for frozen thermal images. Each disc can accommodate up to 40 pictures. Various computer programs have been implemented, primarily in FORTRAN although some in PASCAL, for picture storage and retrieval, picture manipulation and both grey level and temperature analysis.

Results

The DUWIR system provides extremely good imagery. The 8-12 micron picture is comparable to a standard Class II TICM system, whilst the 3-5 micron image is only slightly noisier. On slow moving or static targets a very small amount of video frame to frame integration yields images of comparable signal to noise ratio in both wavebands. This is achieved by the use of recursive addition of the form:-

$$F_n = k.F_n + (1-k).F_{n-1}$$

where F_n is the nth frame (i.e. current frame),

         F_{n-1} is the previous frame,

and $k = 2^{-s}$, where s is the shift factor for which a value of 2 was found to be optimum for best S/N ratio improvement whilst minimising image smear.

Examples of images from each waveband, after recursion, are shown in Figures 9 and 10. The image uniformity and the quality of both temperature and spatial resolution are obvious. It should be emphasized that these images were taken simultaneously, and in real time, from standard television monitors.

The high quality of these thermal images permits straightforward subjective comparison of the target and background characteristics in the two prime thermal wavebands. Analysis of the grey levels in the scene, combined with the reference temperature information, leads to detailed information, at high spatial resolution, on the apparent temperatures in the thermal scene.

An example of this analysis is given in Figure 11, which shows isothermal plots of a Land Rover in the 8-12 micron waveband. More detailed results can be represented by a plot of a single TV line (which incorporates the temperature references), as shown in Figure 12, or as area plots (Figure 13 and 14). Further analysis of data of this type yields a detailed understanding of the characteristics of thermal targets.

One of the parameters of particular interest to those modelling system performance is the apparent target to background temperature difference. This can be simply derived from these calibrated thermal images. One way of presenting this information is given in Figure 15 which shows a histogram of the number of pixels at any one temperature as a function of temperature for the Land Rover target and its surrounding background. The mean of these distributions yield an average target temperature of XX +/- XX and an average background temperature of XX +/- XX. Therefore the apparent temperature difference for this example is XX. Clearly this value will be strongly dependent on metrological conditions, time of day etc. and a comprehensive programme of analysis would be required to fully characterise any one target.

Conclusions

A unique high performance thermal imager has been built by GEC Avionics (Basildon, UK) for RSRE. The system is based on UK common module hardware. It provides simultaneous, synchronous imagery in the 3-5 micron and 8-12 micron thermal wavebands, which when combined with the internal references, enables the apparent temperature of objects in the thermal scene to be evaluated. The additional system features of fully automatic operation and the extensive alphanumerics combine to yield an extremely powerful research tool for scene analysis. It fulfils a long outstanding need and is creating widespread interest both in the UK and overseas.

Copyright Ⓒ Controller, HMSO, London 1986.

References

1. Cuthbertson, G.M. and MacGregor, A.D., IEE Conference Publication 204. 1981.
2. Cuthbertson, G.M., Nato AGARDOGRAPH No. 272. 2-1 to 2-12. 1983.

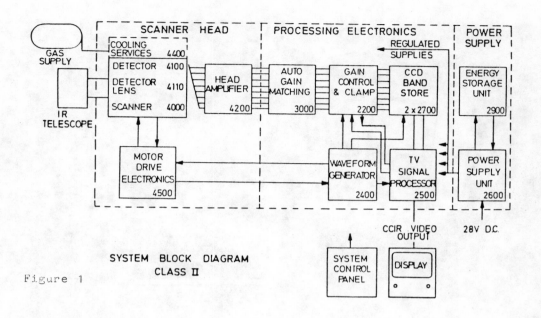

SYSTEM BLOCK DIAGRAM
CLASS II

Figure 1

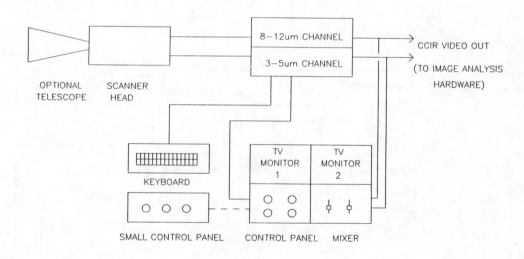

Figure 2 DUWIR system block diagram

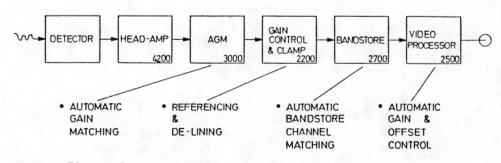

Figure 3 TICM Class II signal processing chain

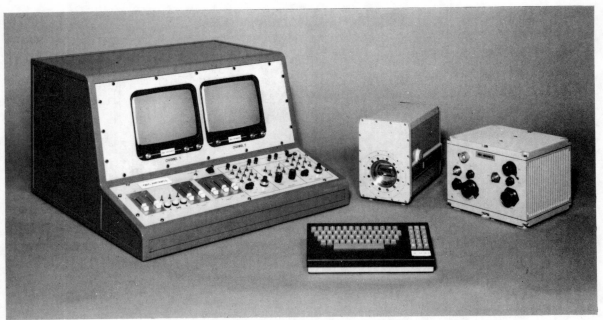

Figure 4   Dual waveband imaging radiometer

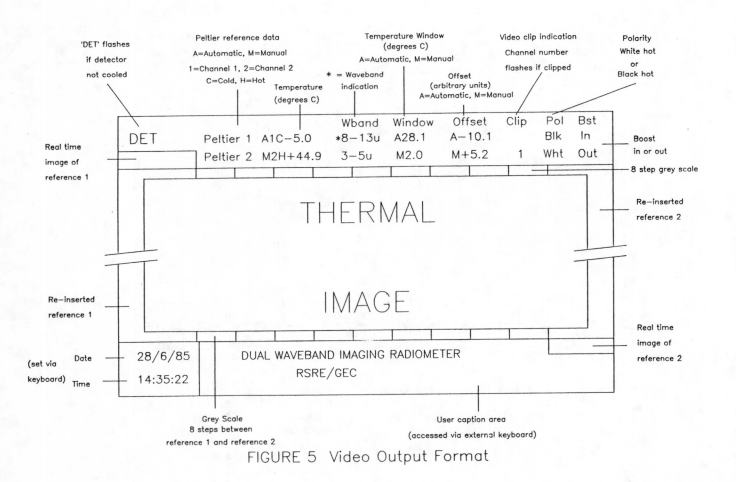

FIGURE 5 Video Output Format

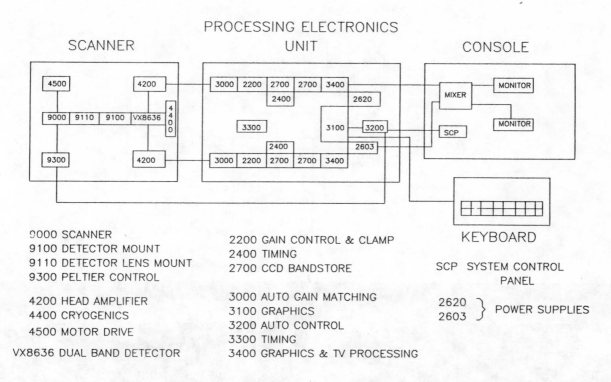

PROCESSING ELECTRONICS

SCANNER   UNIT   CONSOLE

9000 SCANNER
9100 DETECTOR MOUNT
9110 DETECTOR LENS MOUNT
9300 PELTIER CONTROL

4200 HEAD AMPLIFIER
4400 CRYOGENICS
4500 MOTOR DRIVE

VX8636 DUAL BAND DETECTOR

2200 GAIN CONTROL & CLAMP
2400 TIMING
2700 CCD BANDSTORE

3000 AUTO GAIN MATCHING
3100 GRAPHICS
3200 AUTO CONTROL
3300 TIMING
3400 GRAPHICS & TV PROCESSING

SCP SYSTEM CONTROL
PANEL

2620 ⎫
2603 ⎭ POWER SUPPLIES

Figure 6 Configuration of system modules

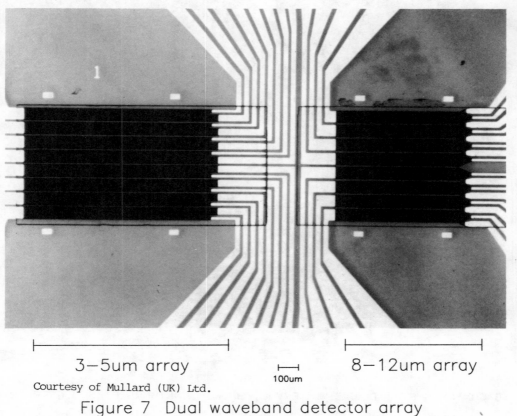

3—5um array   8—12um array

⊢—⊣
100um

Courtesy of Mullard (UK) Ltd.

Figure 7 Dual waveband detector array

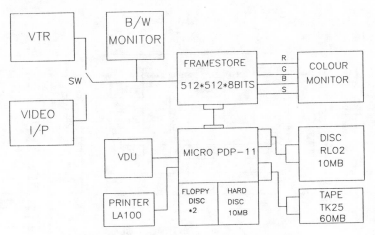

Figure 8   Image analysis hardware

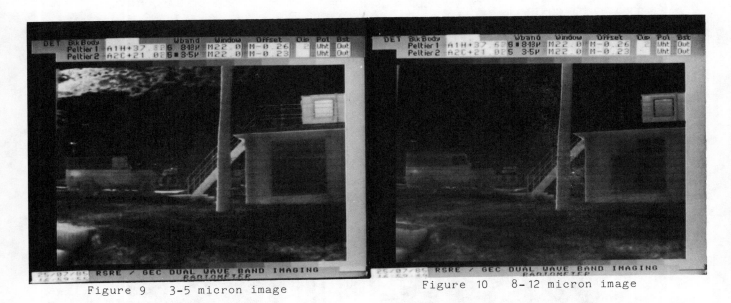

Figure 9   3-5 micron image   Figure 10   8-12 micron image

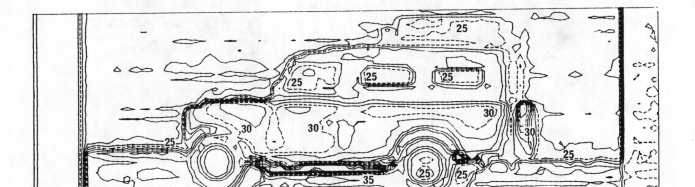

SOLID CONTOURS AT 1 DEGREE INTERVAL, BROKEN CONTOURS AT 5 DEGREE INTERVALS.

Figure 11   Land rover isothermal plot

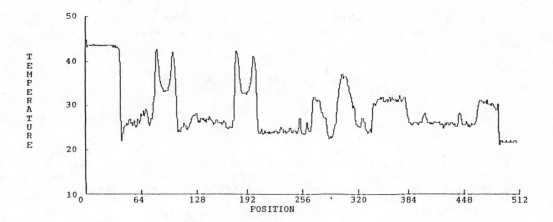

Figure 12   Single TV line temperature profile

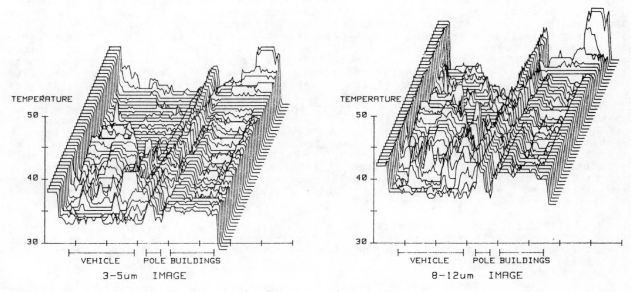

Figure 13   Scene temperature profile / Figure 14   Scene temperature profile

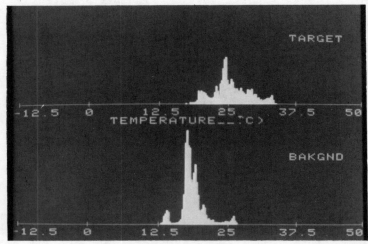

Figure 15   Target and background temperature histograms

Thermal Imaging Sensors for Submarine Periscopes

Herbert M. Runciman

Electro-optics Consultant, Barr & Stroud Ltd,
Caxton Street, Glasgow, Scotland, G13 1HZ, U.K.

Abstract

The evolution in the UK of thermal imaging systems for use in submarines is outlined. A system is described in which a high-performance thermal imager is configured in such a way as to fit within a tubular structure while providing dual-field operation with an elevation facility, giving compatibility with a range of periscope designs. The aspects peculiar to submarine applications are discussed for each part of the system, and a technique for providing rapid surveillance using thermal imaging is described.

Introduction

A key feature of modern defence systems is their ability to operate by day or night. Land forces, in particular, deployed thermal imaging systems at an early stage in their development, since equipment size was not a major consideration, and the equipment could be accessible for maintenance and repair. A prime requirement for submarines, however, is the ability to observe while remaining covert, so that any instrument which significantly increases the exposed volume over that presented by a conventional visual periscope cannot be tolerated. Moreover, due to the positioning of the equipment and the need to provide a pressure seal, little or no maintenance can be carried out, so that until thermal imagers of suitable size and reliability could be produced, reliance was placed on image intensifiers. In 1977 Barr and Stroud constructed the first thermal imager for submarine use with the British Navy, and sea trials took place in 1978. This instrument used a 48 element linear array of CMT detectors which was scanned across the scene in a number of bands by means of a multi-facetted rotor, the picture being generated by a LED array which was viewed from the bottom of the mast via a set of relay optics. For the first trials, a germanium dome with a conventional hard coating was used, and significant attack by sea water occurred. In later trials, a newly developed diamond-like carbon coating was used, and although a significant number of pinholes were present which allowed sea water attack, the dome remains usable after several years of intermittent use.

This prototype adequately demonstrated the feasibility and advantages of providing thermal imaging in a periscope, and led to the development of an imager which was capable of providing a remote TV-based display in addition to an LED display for direct viewing. This system is still in service.

Since then, with the invention of the SPRITE detector at the Royal Signals and Radar Establishment (Malvern) and the development of the high speed scanners necessary to take full advantage of these devices, the performance of thermal imagers has been greatly improved, and equipment size has been reduced, giving great potential benefit for submarine systems. Accordingly, a high performance imaging system for periscope use was designed. The requirements and constraints included the following:-

(i)   A SPRITE-based imager was to be used.
(ii) The system was to be modular to allow maximum flexibility.
(iii) The system was required to fit into as small a tube as possible.
(iv) Dual field operation was required.
(v)   Viewing at high elevation angles had to be provided.
(vi) High reliability and easy maintainability were required.

The resulting system was based on the Barr and Stroud IR18 imager with some alterations to suit use in a periscope. In the following sections the aspects peculiar to submarine use are discussed in the order in which the incoming IR signal passes through the system. The system is shown diagrammatically in fig. 1.

The Germanium Dome

A feature unique to submarine systems is the requirement for a thermally transparent dome or window which is capable of withstanding not only chemical and biological attack by the sea water environment, but also the very high pressures which result when the submarine is at maximum depth. The choice between a plane window and a concentric dome was determined partly by mechanical and partly by optical considerations. A plane window has the great advantage that its optical properties are invariant with the position of the entrance pupil on the window, and no aberrations are introduced into a parallel beam, so that the rotation axis for elevation can be placed arbitrarily, making accurate alignment between the window and the remainder of the system unnecessary. A dome, by contrast, introduces optical power, spherical aberration and chromatic aberration even when the pupil is at the centre of curvature of the dome with coma and astigmatism if the pupil is elsewhere. The elevation axis must thus pass strictly through the centre of the dome, so alignment between the dome and the imager behind it is important. Provided that the latter conditions are met, however, it is found that if high elevation angles are required the diameter required of a dome for a given pupil size is significantly smaller than that required of a plane window. Moreover, for a given thickness and diameter, a dome can withstand much greater pressure than a plane window. A solution based on a dome was therefore selected.

The material chosen was polycrystalline germanium, since adequate spatial resolution is obtained, and the cost of production is significantly lower. The optical properties of germanium are excellent, dispersion and absorption both being much lower than for zinc sulphide, for example. The only serious disadvantage with germanium is the rapidity with which unprotected germanium is attacked by sea water, so that its use is dependent on a highly effective protective coating.

The coating used was ARG4, a diamond-like carbon coating produced by plasma deposition using a proprietary process. The coating has now been in production for over seven years, and its first application to a large component was to a submarine periscope dome. The coating is extremely hard, and has withstood 250,000 wipes with a sand-loaded wiper in laboratory tests. More significantly, it is very resistant to sea water attack. Early versions of the coating suffered from "pinholes", - despite which the domes remain operational after 4 - 5 years of normal use - but a major investment in equipment and development effort has resulted in a process which virtually eliminates pinholes, making germanium a practical material for the sea-water environment even for extended exposure. More recent developments include a multi-layer plasma coating technology which gives the same mechanical and chemical resistance properties as ARG4, but with the added advantage of greater flexibility in design of spectral transmission characteristics. For example, a dual-band coating (ARG6), giving reflectivity less than 4% over the 3 - 5um waveband and less than 5% over the 8-12um band is now in production.

These coatings have been found in practice to allow rapid drainage when the sensor is raised, permitting almost immediate operation of the imager in most cases.

Provision of elevation and azimuth control

It is important that the system should be capable of searching as much of the upper hemisphere as possible. If the field of view of the imager were made sufficiently high to cover the desired elevation, the spatial resolution would be much too low, so steering in azimuth and elevation is required. Azimuth control was provided by rotation of the entire periscope, since otherwise a hyper-hemispherical dome would have been required which would have been costly to manufacture and which would have precluded the fitting of any other equipment or sensors above the TI sensor. There were a number of possible options for elevation control, which included tilting the complete imager, use of an elevating mirror, and use of an elevating prism.

Because of constraints on tube diameter, tilting of the sensor was at once ruled out for this application. The difficulty with the use of a mirror is the size required to give large elevation angles, the problem being exacerbated by the requirement imposed by the optical characteristics of the dome that the rotation axes must pass through the centre of curvature of the dome. If the mirror is placed between the dome and the telescope of the IR sensor, the large mirror size necessitates a long optical path between the dome and the telescope objective, and it is found that the difficulty of compensating the power and aberrations of the dome in the subsequent optics is increased. These difficulties can be reduced at the expense of increased mechanical complexity by placing the elevation mirror behind the telescope objective, and causing the latter to move at twice the speed of the elevating mirror. This solution had been used successfully on previous occasions, but it conflicted with the modular concept, since the elevation control would then become an integral part of the telescope design so that a major alteration would be required if a need for different magnificiations were to arise for use in other periscopes. Also, to reduce development time, it was considered desirable that an IR telescope based on an existing design and mechanically compatible with it should be used, so solutions in which the elevation mechanism had to be included within the telescope were rejected. The possibility of using a refracting prism to provide elevation was examined, the problems in this case being the size of the prism, with attendant cost and optical absorption, and the aberrations introduced. The latter occur because there is no fixed point about which a prism can be rotated which results in the deflected beam having a well-defined elevation pupil over a wide angular range.   The result is that aberrations which are dependent on the the elevation angle are introduced, and these cannot be corrected in the subsequent optics. By optimisation of the position of the prism rotation axis, however, it was found that these aberrations could be kept to an acceptable level for all angles up to zenith.   The prism approach is thus a viable option where extreme elevation angles are required in a limited space, but this is only achieved at the cost of performance.

For the particular application for which the imager was designed it was found that by careful control of pupil positions it was possible to meet the elevation specification by the use of a mirror between the telescope and the dome - the simplest and most direct approach - and this technique was adopted.   The mirror was operated by an angular position servo to give elevation and stabilisation if required.

The IR telescope

The telescope used to convert the scanner field of view to the desired value was based on an existing dual-field short telescope design, re-optimised to take the dome into account. A short telescope was preferred since it reduces the length of tube occupied and thus increases the number of periscope types with which the imager can be used.   (The short length also reduces the required diameter of the lenses for the visual bypass, if fitted). More significantly, the design was chosen because of its desirable pupil characteristics. Because of the long optical path between the objective and the dome it is found that if the entrance pupil lies on the telescope objective the sizes of elevation mirror and dome must be increased, or severe vignetting must be tolerated.   This is particularly serious in the wide-field mode.   With the chosen telescope design it was found to be possible to project the entrance pupil forward so that it lay on the dome, the pupil being defined as the image of the scanner pupil formed by the telescope.   A telescope of this type, showing pupil positions, is shown in fig. 2, and is described in ref. 1.   The scanner used is almost completely free of pupil movements due to the scanning action, so no allowance had to be made for this.   (Movement is less than 1% of beam diameter).

The main requirements for the field of view change mechanism were for reliability and rapid operation, and lenses mounted on a carousel are moved into position using a Geneva mechanism which provides a positive location in both fields of view in a fail-safe manner.

Athermalisation is performed by a position servo which moves an optical element in response to the output of a temperature sensor.   A manual focus control is also provided which can be used to override this servo if desired.

The scanner

At the commencement of the design, it became apparent that no existing scanner of the required performance would fit directly into a tube of the specified diameter without modification.   The Barr and Stroud IR18 scanner was selected partly because it was the smallest of the high performance imagers, and had the lowest power consumption, and partly because (since it is an indigenous product) the highly detailed information necessary to carry out modifications and reconfigurations was available.

The essential components of the scanner are a transfer lens to re-image the detector, a line scan rotor which causes this image to move in a circular arc, a collimator mirror which also forms a real image of the line scan pupil and a frame scanner comprising a mirror driven by a limited-rotation torque motor with a closed-loop servo system.   The mode of operation is described in reference 2.   In its original form, the IR18 scanner also contains fold mirrors to give the optimum "general purpose" configuration, in which the line of sight is horizontal.   For periscope use, it is necessary that the line of sight should be vertical.   Apart from the fact that the imager would not fit into the desired tube diameter, there are disadvantages associated with merely turning the imager so that it points upwards.   Although the attitude of the scanner is not in any way critical, it is desirable that to maximise reliability the detector should be vertical so that the liquid air formed by the cooler remains in contact with the detector substrate, and that the line scan motor should have its axis of rotation lying in a substantially horizontal plane.   The line scan rotor is driven by a self-pumping gas-bearing motor (developed by Ferranti Ltd) which has conical bearings, and if the rotor has its axis vertical there is a tendency for unequal wear on the upper and lower bearings when the motor is subjected to vibration while static, since the gas then provides no cushioning effect.   With the more recent models of motor now in use, however, it is probable that no restriction in attitude is required.   A very long maintenance-free operational life is achieved.   The reliability is aided by the fact that the line scanner is a sealed unit - a scanner of this type has operated without failure for more than 40,000 hours in a military application.

The fold configuration adopted is shown in fig. 3. It will be noted that almost all of the scanner lies to one side of the vertical optical axis. This has the advantage of leaving room for an optical bypass in periscopes which require it, so that the operator has a direct visual channel which is first out of the water.

Detector

The detector is an 8-element SPRITE of which four channels are normally used. The scan speed and detector bias are such that these channels operate at optimum performance, and the small area of the focal plane which is utilised allows nearly perfect cold shielding wihout re-imaging of the exit pupil. The electronics, particularly those required for channel equalisation, are simpler and more reliable than would be required for eight channels. Extension to eight channels, giving a 40% gain in sensitivity, is possible as a future "stretch" option.

Detector Cooling

The provision of reliable detector cooling is a major consideration for periscope systems. The options are the use of a cooling engine or Joule Thomson cooling, since the required temperature of about 80K cannot readily be achieved by thermoelectric means, and techniques involving bulk liquid nitrogen can at once be ruled out. Apart from the size constraints, there are limits on power dissipation and acoustic noise generation within the periscope. Also, extreme reliability is required as access to the sensor head is not possible during a voyage.

The suitability of a number of cooling engines was examined. They have the distinct advantage of requiring only electrical inputs, but they occupy a significant amount of space, generate some acoustic noise, and greatly increase the heat load that must be dissipated within the periscope. Reliability was also somewhat questionable, although engines are improving in this respect. With a split-cycle engine some of these problems could theoretically be eliminated if the parts of the engine could be separated by tens of feet rather than the short distance usually permitted, but this would not help maintainability unless the engine were demountable. Since this would involve loss of the working fluid (usually helium) it would almost certainly not be practical for on-board maintenance. In addition to these considerations, it was found not to be possible to fit the imager with any of the available cooling engines into the space available, so the use of a cooling engine was abandoned, at least for the requirement for which the prototype was designed.

Joule-Thomson cooling was thus adopted in the form of a mini-cooler supplied with compressed air. The provision of the latter required careful consideration. Since regular access to the imager is not possible, a guard drier near the imager can be used only only if the incoming air is sufficiently dry and pure that it has an essentially infinite life. Such air can be supplied either from the submarine systems via gas cleaning apparatus incorporating molecular sieves, or from a dedicated mini-compresser. In the former case, highly purified air is supplied via a flexible high-pressure pipe in a dip loop to a rotary high-pressure coupling in the base of the periscope. The principal reason for the use of a mini-compressor is to remove the need for the flexible high pressure pipe and rotating joint, since it can be attached directly to the periscope. The other important aspect of J-T cooling is the need to provide an escape for the exhausted air, since in the absence of this the pressure within the periscope would gradually increase to an unacceptable value. The supply and exhaust of the air is therefore controlled by a number of solenoid-operated valves in a fail-safe configuration. If the periscope has an optical channel, the change of refractive index of the air with pressure can have a significant effect on focus, so the exhaust valve is controlled to maintain constant absolute pressure within the periscope tube (both ends of which are sealed so that an accident involving the top end of the periscope does not result in flooding of the submarine). The controlled pressure is higher than the maximum expected within the hull, so that exhausting is always possible. Another pressure sensor cuts off the incoming air supply if a pre-set danger level is exceeded. In general the air is exhausted into the submarine directly, but if the mini-compressor is used it is returned to the input of the compressor, prolonging the life of the molecular sieves and drier.

Imager Electronics

The electronic circuitry is almost identical to that of the most recent versions of the IR18 imager, but the physical layout was significantly changed to enhance maintainability. A block diagram is shown in fig. 4. Ideally, as much of the electronics as possible should should be removed to the foot of the periscope so that it can be maintained at sea. In practice a compromise must be reached, since the transmission of low-level signals from the detector would make the system very susceptible to interference from high power AC circuits such as servo or line motor drive currents. Some buffering of the signals is thus required at the scan head. It is also important that a built-in test facility (BITE) should be included at the scan head. To avoid transmission of high-power rapidly varying signals and the attendant interference problems, the drive circuits for the line scan motor and frame scan motor are also located at the scan head, although the timing and control circuits for these motors is placed at the foot of the periscope. To reduce the amount of cabling, and to keep impedances low, a separate power supply module is used for the electronics associated with the scan head, this being located beside the scan head.

At the foot of the periscope, where on-board maintainance can be carried out, is placed the remainder of the imager electronics. This includes the scan converter, which converts the parallel output from the detector to serial form, a microprocessor controlled de-liner circuit to remove the effects of any channel imbalance, a timing and control generator, and a power supply. An interface to allow overlay of alpha-numeric information on the display is also included.

Display and Controls

The output from the imager is directly compatible with standard 525 line 60Hz monitors, so complete flexibility of display configuration is possible. In a typical installation there are two displays - one comprising a small CRT within the ocular box, and one remote display. The display in the ocular box is viewed through the same eyepiece as is used for visual observation, and a means of switching between visual and thermal observation is provided. All imager controls are duplicated at the remote display, and by operation of a switch either set of controls may be made to over-ride the other.

Controls provided include gain (contrast), black level, focus, field of view change and polarity inversion. The black level may be automatic or manual, and may be referenced from the mean scene level or from an adjustable reference level. The advantage of the latter becomes apparent if the scene contains a bright target (such as the sun) which would depress the black level with mean scene referencing.

Optical Bypass

Although it is not strictly part of the thermal imaging system, the optical bypass can be considered to be an optional module which is configured round the imager. The bypass comprises a unit magnification afocal telescope having its entrance pupil and exit pupil brought to coincide with the centre line of the periscope tube by means of prisms. The bypass must of course be designed to suit the particular periscope with which the imager is to be used, or the periscope must be designed to be compatible with an existing bypass design, so the by-pass cannot be treated as a completely independent module.

Maintainability

The system has as far as possible been split into two distinct parts at the top and bottom of the periscope. To minimise the required number of interconnections certain parts (such as power supplies) have been duplicated. Although electrical cables have been used to date, all signals (excluding power supplies) could be transmitted using a bi-directional fibre optics bus. Much of the necessary buffering is included.

Built in test equipment (BITE) is incorporated so that appropriate action may be taken in the event of failure. Indication is provided to a high degree of confidence as to whether failure has occurred in the accessible or the inaccessible part of the equipment. The equipment is divided into modular LRA's (Lowest Replaceable Assemblies), and the BITE system indicates in which of these failure has occurred. Indication to PEC level is achieved with a lower degree of confidence.

Although not accessible at sea, the thermal imaging scan head and its associated optics and electronics are also designed on the LRA principle, and a separate BITE facility for this section is incorporated. Because of the interchangeability of the LRA's, any LRA can be rapidly replaced and the faulty one can be incorporated in a shore-based rig for more detailed testing and repair.

Alternative Configurations

The system described is modularly constructed to suit a wide variety of periscopes with relatively minor modifications. Telescopes to give any desired magnification that is achievable within the specified tube diameter may be fitted, and different trade-off's between maximum elevation angle and field of view can be made according to the user's requirements. Since the scan head is derived from one having a horizontal line of sight, the latter can be used, if space constraints permit, to provide elevation by tilting the scanner/telescope combination, thereby removing the need for an elevating mirror and allowing higher elevation angles. Three different periscope types incorporating this imager system are currently undergoing full development.

Rapid Surveillance

A prime requirement for a submarine is the ability to remain covert at all times, and in order to do this it is desirable to minimise the time for which the periscope protrudes above the sea surface. There is therefore a need to provide rapid all-round surveillance. This can be done by rotating the periscope quickly and recording the thermal imager output on videotape or disc but if this is done the picture is severely distorted if the azimuth rotation speed approaches the frame scan speed, so that electronic correction is essential. Moreover, the elevation coverage in each azimuth sweep is limited to the vertical field of view of the imager. If however, the imager is orientated in such a way that the scan lines are vertical, and the display is similarly orientated, the resulting distortion is merely horizontal compression or expansion, rather than the rhombic distortion occurring when the conventional arrangement is used. This distortion may be removed completely by controlling the amplitude of movement of the frame scan mirror in such a way as to maintain the correct aspect ratio, so that the video recorder receives a series of undistorted "snapshots" which may be examined after the periscope is retracted. The vertical field covered in each sweep is then determined by coverage in the line scan direction, which is some 50% greater than is obtained with the line scan horizontal. If the azimuth sweep speed is suitably chosen, it is possible to switch off the frame scan motor entirely, so that the instrument behaves as a line scanner. In this mode the effects of frame scanner efficiency disappear, so the horizon can be scanned very quickly, although standard video recording is not possible without some loss of information. The penalties to be paid for the rapid gathering of data in this way are that interlace is lost, giving a degradation in horizontal spatial resolution and increased likelihood of aliassing, and that the sensitivity of the system is no longer increased by eye integration from frame to frame. The technique has been demonstrated in a shore-based over-water trial against ship targets using an imager similar to the one described. In practice, all that would be necessary to provide this facility in the periscope would be to rotate the imager through 90° within the tube, leaving the elevating mirror in its original position. This could be done either as a permanent modification, or, if the additional cost and complexity could be tolerated, imager attitude could be remotely controlled. The technique is described in reference 3. Rotation of the display by 90° would normally be accomplished by re-addressing a frame store so that compatibility with remote monitors would be retained. The usefulness of this facility would be further enhanced if automatic target detection were used to present only regions surrounding potential targets on the display, thereby reducing operator fatigue and search time.

Conclusions

Following the successful deployment in submarines of a number of dedicated thermal imagers, a thermal imaging system for submarine periscope use has been developed which, by virtue of its modular construction, not only has a high degree of flexibility and maintainability but is capable of evolutionary improvement and re-configuration in line with future advances in thermal imaging or periscope systems.

Acknowledgements

The author acknowledges with thanks the help he has received from his colleagues within Barr and Stroud Limited in the preparation of this paper. He also thanks the directors of Barr and Stroud Limited and the Ministry of Defence for giving permission for publication.

References

1.   I.A. Neil:- UK Patent GB2152227B "Infra Red Optical System".
2.   P.J. Berry & H.M. Runciman:- UK Patent GB1586099 "Radiation Scanning System".
                (US Patent US4210810)
3.   P.J. Berry & D.S. Ritchie:- UK Patent GB2048606 "An Optical System".
                (US Patent US4246612)

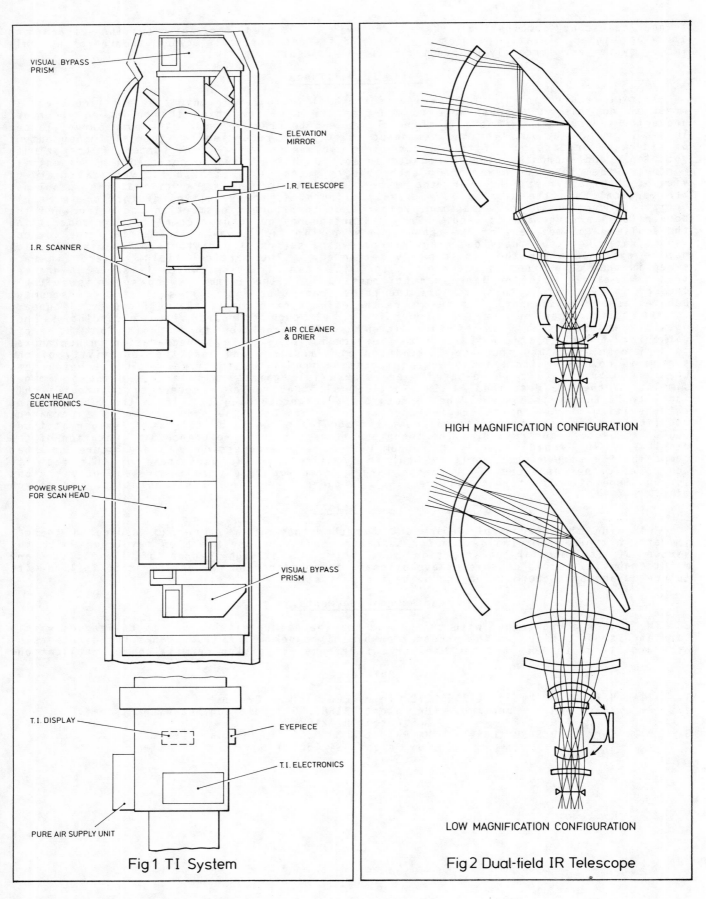

VISUAL BYPASS
PRISM

ELEVATION
MIRROR

I.R. TELESCOPE

I.R. SCANNER

AIR CLEANER
& DRIER

SCAN HEAD
ELECTRONICS

POWER SUPPLY
FOR SCAN HEAD

VISUAL BYPASS
PRISM

T.I. DISPLAY

EYEPIECE

T.I. ELECTRONICS

PURE AIR SUPPLY UNIT

Fig 1 TI System

HIGH MAGNIFICATION CONFIGURATION

LOW MAGNIFICATION CONFIGURATION

Fig 2 Dual-field IR Telescope

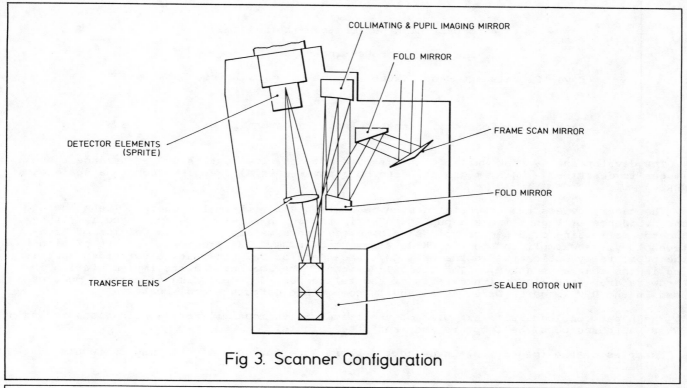

Fig 3. Scanner Configuration

Labels in figure:
- COLLIMATING & PUPIL IMAGING MIRROR
- FOLD MIRROR
- FRAME SCAN MIRROR
- FOLD MIRROR
- SEALED ROTOR UNIT
- DETECTOR ELEMENTS (SPRITE)
- TRANSFER LENS

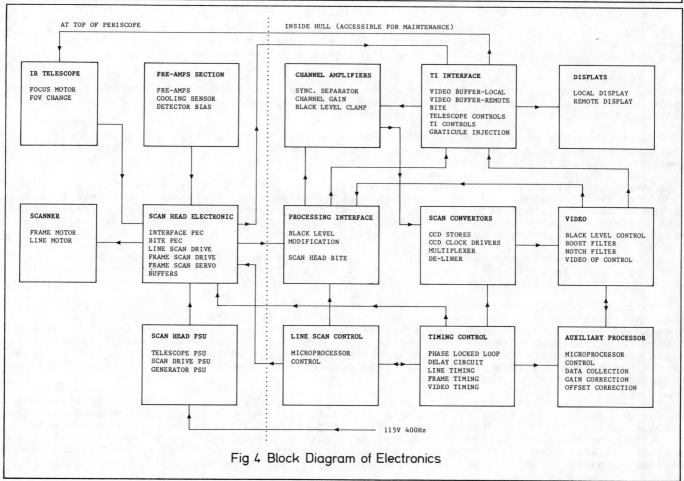

Fig 4 Block Diagram of Electronics

AT TOP OF PERISCOPE   INSIDE HULL (ACCESSIBLE FOR MAINTENANCE)

IR TELESCOPE
FOCUS MOTOR
FOV CHANGE

PRE-AMPS SECTION
PRE-AMPS
COOLING SENSOR
DETECTOR BIAS

CHANNEL AMPLIFIERS
SYNC. SEPARATOR
CHANNEL GAIN
BLACK LEVEL CLAMP

TI INTERFACE
VIDEO BUFFER-LOCAL
VIDEO BUFFER-REMOTE
BITE
TELESCOPE CONTROLS
TI CONTROLS
GRATICULE INJECTION

DISPLAYS
LOCAL DISPLAY
REMOTE DISPLAY

SCANNER
FRAME MOTOR
LINE MOTOR

SCAN HEAD ELECTRONIC
INTERFACE PEC
BITE PEC
LINE SCAN DRIVE
FRAME SCAN DRIVE
FRAME SCAN SERVO
BUFFERS

PROCESSING INTERFACE
BLACK LEVEL
MODIFICATION
SCAN HEAD BITE

SCAN CONVERTORS
CCD STORES
CCD CLOCK DRIVERS
MULTIPLEXER
DE-LINER

VIDEO
BLACK LEVEL CONTROL
BOOST FILTER
NOTCH FILTER
VIDEO OP CONTROL

SCAN HEAD PSU
TELESCOPE PSU
SCAN DRIVE PSU
GENERATOR PSU

LINE SCAN CONTROL
MICROPROCESSOR
CONTROL

TIMING CONTROL
PHASE LOCKED LOOP
DELAY CIRCUIT
LINE TIMING
FRAME TIMING
VIDEO TIMING

AUXILIARY PROCESSOR
MICROPROCESSOR
CONTROL
DATA COLLECTION
GAIN CORRECTION
OFFSET CORRECTION

115V 400Hz

A compact high performance thermal imager

A H Lettington

Royal Signals and Radar Establishment Malvern, Worcs, UK WR14 3PS

W T Moore

Rank Pullin Controls, Langston Road, Laughton, Essex, UK IG10 3TW

The development by Rank Pullin Controls of a compact, low cost, high performance thermal imager is described. This imager is based on a novel coaxial scanning technique designed at RSRE.

The paper reviews the essential design requirements for thermal imagers. These include the avoidance of Vignetting and pupil wander in the scanning elements and objective lenses, and the maximising of detector cold shielding. These aims are met by (a) choice of array technology, as exemplified in the UK development of SPRITE IR detectors, (b) use of either the pupil relay optics (as in the UK TICM II Scanner) or the alternative coincident pupils (as in the coaxial scanner). The optical design of the scanner and its implementation in hardware are given along with details of the telescopes, detectors and electronic modules used in the RPC imager. Direct View and TV compatible outputs are discussed.

The paper concludes with examples of the imager in various military applications including RPV's, airborne pods, helicopters and land based sights.

Examples of the imagery obtained and the thermal performance of the imager are also described.

Introduction

In this paper we describe the design and development of a miniature high performance thermal imager. It was intended originally for a specific RPV requirement but has now attained more general acceptance particularly where size, weight and cost are at a premium.

There are a number of constraints which affect the design of any compact thermal imaging system. For example there is the need to maximise the Dθ product of the scanner with respect to its physical size. It is also generally desirable to avoid vignetting and pupil wander and to maximise the detector cold shielding.

A scanner with a high Dθ product and small size inevitably requires a high scan angle. This must be used in conjunction with an afocal telescope to match the scan angle and aperture required at the objective lens with those of the scanner.

A small high speed rotating polygon with reflective facets usually forms the basis of such a compact afocal scanner. For a given diameter of polygon the Dθ product increases inversely as the square of the number of facets. It is difficult however to design afocal telescopes with angles at their exit pupils of greater than 60 degrees. This immediately governs the diameter of the rotating polygon and indirectly the number of facets, which is typically 6.

Six swathes however are not usually sufficient in a high performance afocal imager and a flapping mirror is frequently used in conjunction with the rotating polygon to achieve scans in both azimuth and elevation.

The desire to maximise the detector cold shielding is best met using a matrix of detector elements closely packed into a square active area. The UK SPRITE IR detector[1] is an example of such an array which also achieves the signal processing of time delay and integrate within the serial dimension of the detector element. By using this square format the pupil wander associated with the angular variation from top to bottom and from left to right of the detector array is minimised. It also enables the detector to be tightly cold shielded so improving its thermal sensitivity.

The scanning mechanism can be the largest source of pupil wander. If the cold shielding at the detector is not to be relaxed or additional rotating cold shields introduced then the various optical components including the front objective must be overdimensioned to avoid vignetting at the edges of optical components. When vignetting occurs or when the detector aperture is increased to accommodate pupil wander severe cosmetic defects, such as picture shading, usually occur in the displayed image of the thermal scene.

Pupil wander within the scanner can be minimised by relaying the pupils through the telescope to the scanner and through the scanner to the detector. An example of an imager employing pupil relay has been described by Moore and Reeve.[2] In this design which has formed the basis of the scanner in the UK Common Module (TICM II) the exit pupil of the afocal telescope is incident on a facet of the rotating polygon and is relayed by the concave mirror onto the flapping mirror. The aperture stop of the detector is close to this mirror so reducing any possible wander associated with the physical dimensions of the detector array.

It is possible to eliminate the relay mechanism between the polygon and mirror if this mirror is over dimensioned and performs lateral as well as angular motion. This introduces additional mechanical complexity and since the flapping mirror is frequently the least reliable of the scanning mechanisms, this approach is thought to be undesirable. It was against this background that the coaxial scanner was designed and developed to produce a compact, high performance thermal imager.

Optical design of the coaxial scanner

The coaxial scanner which was designed within RSRE[3] combines two orthogonal scanning mechanisms into virtually a single rotary motion possessing good radiometry and minimum pupil wander without the need of additional relay optics. This has resulted in a miniature imager with a high Dθ product and optical transmission and with minimum optical distortion. This is particularly so for the scan in elevation where the linearity is set by the accuracy of the facet angles and not by the response of a flapping mirror.

The scanning mechanism is illustrated in Figure 1. It employs two rotating polygons mounted one above the other on the same axis and rotating in the same direction at slightly different speeds. The ratio of their speeds being inversely proportional to the ratio of the number of facets on each polygon. These facets are variously inclined and the combination provides the scan both in azimuth and elevation. The elevation scan is set by the combination of the facet angles that are coincident at any time on the upper and lower polygons and the scan in azimuth is achieved through the rotation of each pair of facets.

The facets on the upper and lower polygons are aligned in the direction which views the infrared scene. Radiation is reflected from the lower polygon onto the upper one and thence to the detector. The coincident facets are accurately aligned at the centre of the horizontal scan but slip in phase as they rotate. In this way it is possible to permutate the inclinations on the lower polygon with those on the upper polygon and achieve a wide scan in elevation.

The particular design which has proved most successful employs 7 facets on the lower polygon and 8 on the upper. This combination can provide 56 equally spaced scans in elevation with a 100% elevation scan efficiency. In this respect the coaxial scanner is similar to a 56 sided polygon with variously inclined facets. The physical size of the coaxial scanner is however considerably smaller than such a multifacetted polygon and the Dθ product considerably greater. Early versions of the coaxial scanner adopted the 100% scan format and used a dedicated monitor. The present range of modules however employs a scan sequence of lower efficiency but that can be used with either a dedicated monitor or a CCIR compatible TV. Details of these modules are given in the following section.

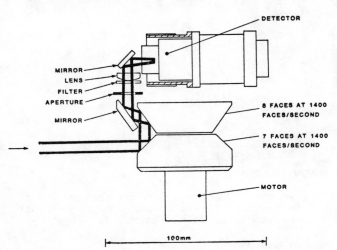

Figure 1. Coaxial scanner mechanism.

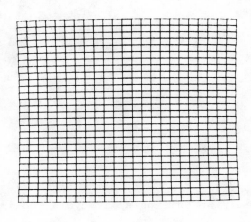

Figure 2. Scan pattern for coaxial scanner.

It can be seen from figure 1 that the radiation is reflected from the upper polygon at an angle to the axis of rotation and that the 8 sided polygon is uppermost. This compensates for any effect of the differing speeds of rotation of the two polygons on the scan pattern of the imager.

It ensures that the horizontal scanning is linear in the centre of the vertical field of view. The resulting scan pattern of the imager is shown in figure 2. This is plotted linearly in angle both in azimuth and elevation and without a telescope covering a field of view of 40° x 30°. In figure 2 each rectangle represents the angular subtense of the full detector array mapped onto the field of view. A more detailed discussion on mapping is given in a later section on Display and Control.

The scanner was designed to have a pupil size of 10.4 mm. This pupil is located at the apex of the roof edge reflector formed by the upper and lower facets and is coincident for both azimuth and elevation scans so minimising pupil wander.

Modular form of the imager

The RPC imager consists of a suite of modules. They are configured differently according to the required application and fall into four main groups - 1) The scan head, 2) Signal processing, 3) Image processing and 4) Display and control.

The definition of these groups has been determined largely by those configurations which required a physical separation of one group of modules from another. These groups are described in the following sections.

The scan head

The modules in the scan head are a) the telescope, b) the scanner, c) the detector array and cryogenics, d) the head amplifier and e) the motor drive circuits.

A wide range of afocal telescopes is available and others are under development. Due to the very low 'pupil scan' of the scanner mentioned previously these telescopes can make efficient use of the objective area without the risk of introducing radiometric errors, thereby combining energy gathering efficiency with the avoidance of picture shading effects.

The scanner module uses the coaxial scanner design described in the previous section. This module includes the drive motor, and polygon phase pickoffs, and is illustrated in figure 3 where its small size is clearly seen.

Figure 3. Coaxial scanner module.

The detector array is a Mullard 8-element SPRITE array, developed as a common module of the UK Thermal Imaging Common Modules programme. The cryogenics are usually high pressure gas, but the Philips split cycle cooling engine is available as an alternative.

The head amplifier array is another UK TICM common module.

The motor drive unit drives the motor, controls its phase and generates synchronisation pulses to track residual motor phase wander.

Signal conditioning

The signal conditioning group consists of four modules. These are a) the analogue module, b) the band conversion module, c) the auto control module and d) the power supply module.

The analogue module performs the function of amplifying the eight signals from the head amplifiers module, applying electronic correction and bandfiltering, limiting the dynamic range of the signal, and removing inter channel offsets.

The band conversion module digitises the eight signals, uses RAM chips to convert the eight parallel signals to a sequential signal suitable for display on a dedicated monitor or for tape recording.

The autocontrol module is microprocessor based. It automates the settings of the thermal imager, provides a graticule injection facility, and offers (if required) signal transfer curve correction. This module was originally intended as an option, but it is our experience that the automatic facility consistently sets the imager up better than the operator, and that this module is highly desirable. A serial port on the module relays the imager settings to other interested units, such as the image processor. Manual override is available although seldom used.

The power supply module powers both scan head and signal conditioning modules from raw 19-32 Volts DC.

Scan sequence

The laws of arithmetic militate against the facet combinations of the coaxial scanner scanning the scene in the exact line order of standard TV. Therefore, one must either display via a dedicated monitor, or via a line resequencing module. In the latter case, however, care must be exercised to avoid cosmetic defects on moving objects or moving scenes (motion tear). Consequently, we take advantage of the fact that it is possible to design scanners with (n:n-1) facets (n even) in such a manner that the scan proceeds progressively within any one field, but alternates between two interlace fields as it does so. Scanners designed in this way do not suffer from motion defects, and offer $\frac{1}{2}(n+2)^2 - 6$ useful swathes per frame. With an eight element detector, this gives 352 scan lines from an 8:7 faceted scanner. The required motor drive power to overcome air drag (for a given radiation gathering efficiency) is approximately proportional to the ninth power of the number of facets, which is another reason for preferring 8:7 facets rather than 10:9. Also the vertical scan efficiency which is 0.79 for n=8 falls as n increases.

Image processing

At present, the group of image processing modules comprises the following:- a) Line resequencing module, b) Frame integration noise reduction submodule, c) Video insertion module and d) Analogue interface module.

The line resequencing module has several functions. Prime amongst these is resequencing the scan lines to standard TV order, thereby obviating the need for dedicated monitors. A second function is aspect ratio correction; the 352 IR scan lines are displayed as 484 TV lines. Interestingly, this increase in the number of displayed lines produces a subjective gain in picture sharpness, even though no more information is actually displayed. The module also provides some scan mapping correction, working to a tenth of a pixel. The module software is being extended to provide freeze frame and electronic zoom facilities.

The optional frame integration noise reduction submodule takes advantage of spare data rate capacity in the resequencing module. It includes a motion detector which suppresses integration in regions of movement. This avoids any blurring due to motion. There is a software facility for automatic selection of integration level and motion threshold as a function of the autocontrol module gain settings. This is part of the philosophy of optimising as nearly as possible the imager function with the minimum of human intervention.

The video insertion module is still under development. It will be fully bit mapped, allowing text or detailed graticule graphics to be overlaid on the picture, with a variety of modes.

The analogue interface module is required when the image processing is performed in a unit remote from the signal conditioning, for example when an analogue RPV downlink is involved.

Additional modules, such as a video tracker are planned.

Display and control

Various display options are available, from a rugged 9 inch monitor to a monocular viewer. These are available in both dedicated and CCIR versions. The dedicated version includes its own mapping correction circuits, since it does not include an image processing unit.

Mapping analysis. Most afocal thermal scanners suffer from an intrinsic mapping error which is anamorphic and, therefore, not correctable by control of telescope distortion. The formula for the mapping from display co-ordinates to a tangent plane at the scene is

$$(\theta, \phi) \rightarrow (\tan \theta, \sec \theta \tan \phi)$$

where θ and ϕ are the nominal scanner azimuth and elevation scan angles. However, a more relevant mapping, particularly for head-up display conditions, is the polar mapping

$$(\theta, \phi) \rightarrow (\tan \theta, \sec \theta \tan \phi) \, U \cot U$$

where $\tan^2 U = \tan^2 \theta + \tan^2 \phi + \tan^2 \theta \tan^2 \phi$

A third order expansion of this gives $(\theta, \phi) \rightarrow (\phi(1 - \frac{1}{3} \theta^2), \theta(1 + \frac{1}{6} \phi^2))$.

The terms $- \frac{1}{3} \phi \theta^2$ and $\frac{1}{6} \theta \phi^2$ represent anamorphic distortion of the scene.

The coaxial scanner is, however, an exception to this distortion analysis. In an idealised coaxial imager in which both polygons rotate at equal rate, the mapping is

$$(\theta, \phi) \rightarrow (\phi(1 - \frac{1}{12} \theta^2), \theta(1 + \frac{1}{24} \phi^2)).$$

In practice the performance is slightly different from this, (a) because the polygons rotate at slightly different rates, and (b) because we electronically are able to remove the $\phi \theta^2$ term.

By computer ray tracing the distortion corresponds to $(\theta, \phi) \rightarrow (\phi, \theta(1 + \frac{1}{12} \phi^2))$.

This distortion corresponds to approximately ½% at the picture corners, and is unusually low for a high performance thermal imager.

Specification of the coaxial RPC imager

The specification of the coaxial imager is shown in Table 1. It operates in the 8-14 µm IR waveband. Examples of typical thermal scenes are shown in Figure 4.

Table 1. Specification of Coaxial Imager

Scanner pupil diameter	10.4 mm	Projected Detector subtense (with x M Telescope)	2.1 mrads (2.1/M mrads)
IR Line	352	Pixels per line	512
Display Lines	484	Swathe Conversion technique	Digital
Basic Field of View	40 x 25 degrees	Sensitivity	
Detector	Mullard 8 element SPRITE	(without noise reduction)	0.15°C
Cooling	Compressed gas or engine	(with noise reduction)	0.05°C

Applications

A wide range of configurations are available. These currently include RPVs[4], helicopters, fixed wing aircraft, land vehicles, security surveillance and missile guidance. A particular advantage for several of these applications is the combination of high performance with a small scan head package suitable for gimballed mounting in a ball.

The scan head package can be small because a) the scanning optics are small and 'pupil efficient', and b) the use of SPRITE detectors makes relatively few electronic channels necessary.

A view of a coaxial scanner mounted in a demonstrator package for the M.L. Aviation Limited Rotary Winged SPRITE RPV is shown in Figure 5.

Figure 4.   Example of Thermal Imagery.

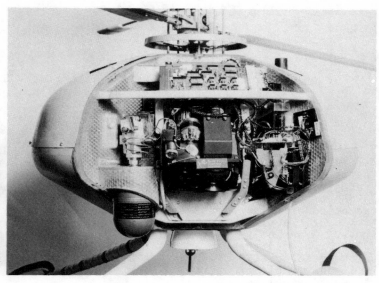

Figure 5.   RPV Package.

References

1. Elliott, C.T., Electronics Letters, Vol. 17, pp. 312-313, 1981.
2. Moore, W.T., Reeve, T.C., Second International Conference on Low Light and Thermal Imaging, IEE Conference Publication No. 173, pp. 63-64, 1979.
3. Lettington, A.H., RSRE Memorandum No. 3292, 1980.
4. Moore, W.T., Woolls, T.D., Fifth International Conference on Remotely Piloted Vehicles, Paper 14.1-14.10, Bristol, 1985.

A Signal Processor for Focal Plane Detector Arrays

R. Kennedy McEwen

Future Systems Laboratory, GEC Avionics Limited
Christopher Martin Road, Basildon, Essex, England.   SS14 3EL

Introduction

This paper discusses a signal processing unit developed for 2D Cadmium Mercury Telluride - Silicon hybrid (CMT-Si) photovoltaic detector arrays. It overcomes the severe fixed pattern noise problems associated with these detectors by performing the required nonuniformity removal in the analogue domain without the use of Digital Signal Processing (DSP) or fast, high resolution Analogue to Digital Convertors (ADC's). The reasons for the adoption of this approach stem from the need for a compact, high speed IR imaging system for use in automatic weapon guidance and homing systems particularly in the anti-armour role. A prototype system using this approach has been developed and used to operate a 32 x 32 element Random Access Line Scanned Array (RALSA) at frame rates in excess of 200Hz with measured NETD of less than 0.1°C. This same system, without further modification, will be capable of operating 64 x 64 element detector arrays with frame rates in excess of 400Hz if required.

Background

Staring IR imaging systems offer many significant advantages over their opto-mechanically scanned counterparts for the weapon guidance application. These include compactness, lower system complexity and ultimately, lower system cost. In particular, as each detector in the focal plane array interrogates a specific area of the scene continuously, the thermal resolution of the system is limited only by the sensitivity of the detector elements, the imager frame rate, and the storage capacity of the integration electronics. The high quantum efficiency of CMT, together with its ability to operate in the 8-12µm band permits the thermal resolution of systems based on CMT-Si detector arrays to be superior to any other form of IR imaging system. Such imagers are ideal for use in missile homing heads and Figure 1 shows a prototype sensor head for such a system.

Currently, staring array detectors are available in a wide variety of formats and materials, however, irrespective of the fabrication techniques and materials involved, all of these detectors suffer from some level of inter-element nonuniformity, or fixed pattern noise. This is present in all focal plane structures and is the result of limitations in the manufacturing processes, but is a particular problem in the IR where scene contrast is typically less than 1 or 2 percent. The importance of some nonuniformity processing system is therefore apparent, as the uncorrected image from these detectors is virtually unrecognisable.

The Non Uniformity Processing (NUP) Algorithm

If every element in the array is essentially linear in response over the temperature window of interest, the output voltage of the ith element can be given by:

$$V_i = R_i T + B_i \qquad (1)$$

R_i = gain of element i

B_i = offset of element i

T = apparent scene temperature

Similarly, the mean output of the array while imaging a uniform scene can be given by:

$$\overline{V} = \overline{R} T + \overline{B} \qquad (2)$$

$$\text{thus} \quad \overline{V} = \frac{\overline{R}}{R_i} \left(V_i - B_i + \frac{\overline{B} R_i}{\overline{R}} \right) \qquad (3)$$

$$\text{or} \quad \overline{V} = K2_i (V_i - K1_i) \tag{4}$$

$$\text{where} \quad K1_i = B_i - \frac{\overline{B} \; R_i}{\overline{R}}$$

$$\text{and} \quad K2_i = \frac{\overline{R}}{R_i}$$

Thus the response of each element in the array may be normalised to the mean response of the array by a suitable choice of $K1_i$ and $K2_i$ for that element. These coefficients can be determined for each element by imaging, in turn, two uniform scenes of temperature T1 and T2 solving the n pairs of simultaneous equations:

$$V_i 1 = R_i \; T1 + B_i$$

$$V_i 2 = R_i \; T2 + B_i$$

to yield B_i, R_i and hence compute $\overline{B}$ and $\overline{R}$

$K1_i$ and $K2_i$ are thus unique for each element and are independent of the reference temperatures T1 and T2. Similarly, the final normalised output of each element given by equation 4 is dependent only on the average response of the array and is independent of R_i and B_i.

Digital Non Uniformity Processing

Due to the extensive use of simple computations in the NUP algorithm the process appears to be ideally suited to implementation in digital hardware. In such systems, calculation of the normalised signal is usually left to dedicated hardware for processing speed, whilst the computation of the correction coefficients is carried out by microprocessor.

Due to the low scene contrast in the IR bands and the level of nonuniformity present in current detector arrays, the signal from the detector must be quantised to at least 12 bits prior to processing, if optimum quality video is to be achieved. For the compact systems proposed for missile homing heads this presents a serious problem as current monolithic 12 bit ADC's are not fast enough to meet the required data rates from the sensor. Currently the maximum data rates available from these devices is around 40kHz, whilst for hybrid convertors, with the necessary sample and hold circuits, data rates as high as 1MHz may be achieved.

As the 2D detector array technology develops, the maximum frame rate of the more compact digital non-uniformity processing units will be extremely limited by the achievable convertor speeds. This will in turn place greater demands on the tracking algorithms of a weapon guidance system.

Analogue Non Uniformity Processing

Although at first sight the implementation of the NUP algorithm in the analogue domain does not appear to be ideal, several benefits of the approach, particularly regarding data rates and dynamic range soon become apparent.

As outlined previously, the signal from the focal plane array must be processed in a system with a dynamic range of at least 12 bits. This implies the use of very low noise circuitry throughout the processor; however this should present no more of a problem than the circuitry used to amplify and sample the signal at the front end, which must be just as noise free, with a similar dynamic range.

Equation 4 shows that the NUP algorithm is simply the subtraction of a background coefficient term $K1_i$ followed by the multiplication by a gain correction coefficient term $K2_i$.

In order to calculate these coefficients, the outputs of each element must be stored when two uniform scenes are viewed. To simplify this process in the analogue system a variation of equation 3 is used:

$$\overline{V} - \overline{B} = \frac{\overline{R}}{R_i} (V_i - B_i) \tag{5}$$

Hence the signal from each element is normalised to the mean output from the entire array less the mean background offset. Naturally this mean offset can be defined at any arbitrary scene temperature and results in that temperature being defined as the initial black level. In practice this restriction is unimportant as the signal can be DC restored after full normalisation to regain the independence on temperature.

Figure 2 shows a block diagram of the prototype system constructed to implement the NUP algorithm in analogue form. Following buffering and pedestal removal, the background offset contribution is subtracted from the signal at the current summing junction A. The coefficient, which is stored in 12 bit RAM (or ROM, should the performance of the sensor be consistent enough) is first converted into analogue form in DAC1.

Following this process, the resultant signal is amplified and fed, as the reference current, to DAC2. Simultaneously, the gain correction coefficient is retrieved from memory and fed to the digital input of the device. The final resultant is the uniform unquantised analogue signal which merely requires sampling to remove any switching pulses before being further processed or fed to a display. If a totally 'hands-off' operation of the unit is required, a simple level shift and overall gain adjustment circuit may be activated in the form of the AGC/ALC shown in Figure 2.

Calibration of the system and calculation of the correction coefficients are performed in the prototype unit by microprocessor, however the routine may be easily implemented in dedicated hardware if speed of calibration is essential.

Initially a uniform cold scene is imaged onto the detector (e.g. a diffuse reflection of itself). The contents of the memory M1 are then successively approximated until the reference current fed to DAC2 is nulled. This effectively computes the term B_i in equation 5, and the signal presented to DAC2 is therefore the output of each element less the background coefficient.

A uniform warm scene is then imaged onto the detector, and the mean output of DAC2 is determined. The contents of memory M2 are then successively approximated until the output of DAC2 corresponding to each element is equal to the previously calculated mean signal. This effectively computes the term $\bar{R}/R_i$ of equation 5 for each element in the detector, and compensates for any change in the responsivity from element to element. Following this initial system calibration, normal scenes may be imaged, with the circuitry of Figure 2 calculating the unquantised resultant of equation 5 in real time.

Figure 3 is a schematic indicating the use of the analogue NUP in a demonstration system already constructed with conventional CCIR video output. The output of the NUP unit is fed directly to an 8 bit video rate ADC to facilitate scan conversion and automatic target tracking algorithms. In this system, a 32 x 32 element RALSA detector is operated at a relatively sedate 200Hz frame rate. Figure 4 shows an image taken directly from this video output stage prior to any processing. The effect of the unit can be clearly seen in the same scene after processing in Figure 5. A typical vehicle is shown in Figure 6. In this image several residual elements can be observed, which have no response at all. No attempt has been made in the prototype system to remove these defects, as a knowledge of their locations in the image is sufficient for the tracking algorithms to ignore them. For an imaging system for human observation, these defects can easily be cosmetically masked by the mean signal from the nearest neighbours.

As shown in the schematic of Figure 3, the tracking algorithms of the prototype system have access to the data in the frame stores which are updated at sensor rate, but the TV frame stores operate at the conventional 25Hz frame rate. One frame of sensor information is grabbed every TV frame, resulting in only one eighth of the sensor information being displayed on the VDU. For the application that the system was designed, this was unimportant, however for an imaging unit for human observation, some form of frame to frame averaging would be desirable. Alternatively, if the video output were the primary output of the system, the sensor could be operated at TV frame rates.

Advantages of the Analogue Approach

Several advantages exist for the analogue NUP, the most obvious being processing speed. As no signal quantisation is involved in the unit, there is no requirement for an ultra-fast, high resolution ADC. The resolution of any subsequent quantisation process (if necessary at all) can be determined by the application rather than by the limitations of the NUP itself. The two monolithic DAC's utilised in the prototype system operate with settling times of less than 200nS at 12 bits, permitting data rates of up to 5MHz without sacrificing final image quality and even faster devices are available. This is considerably faster than any monolithic or hybrid 12 bit ADC's, although fast 12 bit convertor PCB's are available if their size and power dissipation can be tolerated.

In terms of size, the analogue system offers further advantages. The original prototype unit, constructed using only monolithic, off the shelf components occupies less than half of a single size eurocard PCB. Using hybrid construction techniques for development purposes, this volume could be reduced significantly.

In a digital system, the signal from the detector, together with any background components must be limited to the maximum ADC input level if "clipping" of the output is to be prevented. For the analogue unit, the background component itself is quantised to 12 bit resolution for storage purposes - a greater resolution than can sensibly be achieved using a digital system - whilst the peak signal level is limited only by the saturation limit of the front end amplifiers themselves. Some advantage in signal dynamic range is therefore achievable by this system.

Finally, the implementation of automatic gain and automatic level control, as shown in Figure 2 is extremely simple to implement in the analogue domain, permitting hands-off operation to be built into any system.

<u>Conclusion</u>

A non uniformity processing (NUP) system for CMT and similar detector arrays, in which the fixed pattern noise is removed while the signal is still in the analogue domain has been successfully demonstrated. The unit offers significant advantages over conventional DSP techniques, particularly in signal handling capability and bandwidth as it is not limited by current ADC technology. The compact processor is therefore ideally suited to applications requiring relatively large detector arrays, and particularly in homing heads where high frame rates may be essential.

Figure 1. A Prototype Staring Array Sensor Head

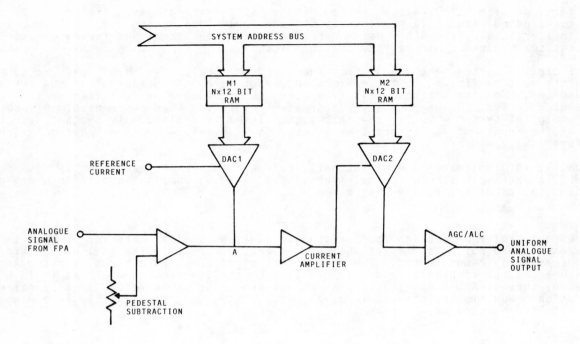

Figure 2.   Schematic of the Analogue Non Uniformity Processor

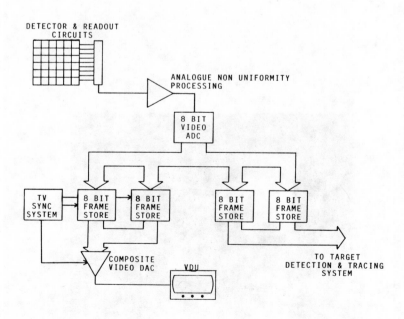

Figure 3.   Schematic of a Demonstration System Utilising the Analogue NUP Unit

Figure 4. Image Before Processing

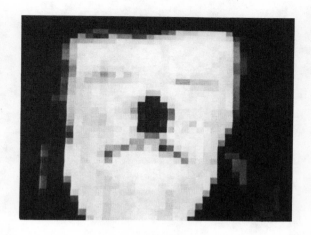

Figure 5. Same Image After Processing

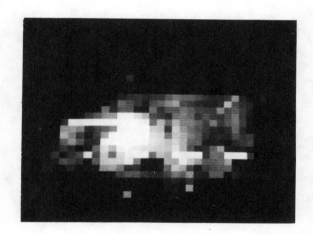

Figure 6. Image of Land Rover After Processing but Leaving 'Dead' Elements

INFRARED TECHNOLOGY XII

Volume 685

Session 4

Infrared Applications

Chair
Richard A. Mollicone
Analytic Decisions Incorporated

Space-Variant PSF Model of a Spirally-Scanning IR System

Timothy G. Bates
Sandia National Laboratories
Division 324
P. O. Box 5800
Albuquerque, NM 87185

Michael K. Giles
Department of Electrical and Computer Engineering
New Mexico State University
Box 3-0
Las Cruces, NM 88003

Abstract

Linear systems theory combined with Fourier transform techniques provides a powerful means to characterize electro-optical systems. We present the description and analysis of a particular scanning IR system whose characteristic Point Spread Function (PSF) is space-variant. This peculiarity of the PSF arises from the scanning geometry; the IR sensor spins as it drops toward the ground.

A system impulse response, or Point Spread Function (PSF), is derived from effects including diffraction, motion blur, chromatic aberration and detector area. Other effects may be taken into account by cascading PSF's. Once the system PSF is determined, the output of the system is the convolution of the input irradiance with the PSF. Alternatively, the output is the familiar inverse Fourier transform of the product of transforms. However, the space-variance of the PSF makes the latter option impossible and the former difficult. The convolution progresses in a spiral manner, with the PSF width changing continuously.

The method is tested using Fortran programs and a digital image processor. The digital images produced show expected results, revealing a marked increase in resolution as the sensor nears the ground.

Introduction

The particular IR system modeled in this paper is one which is dropped from several hundred feet above the ground and spins at a constant rate as it also drops with constant velocity. The detector is square, but the footprint of the detector on the ground is not square since the field of view is centered at an angle of depression. The exact optical system is unspecified; it is sufficient to know only the exit pupil diameter and its distance from the detector. From Figure 1, it is easy to imagine that the footprint will trace out a spiral as the system drops and spins. Note that the footprint area continually decreases.

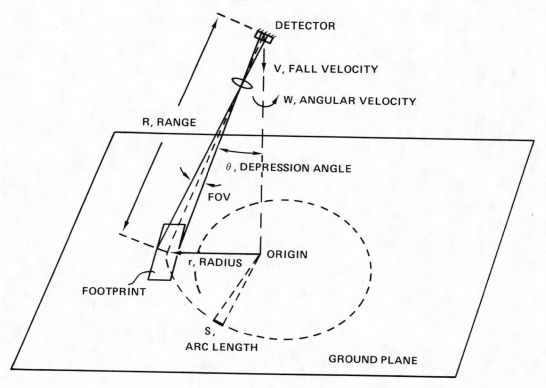

Figure 1. System Geometry

The system point spread function

Linear systems theory is a useful tool for analyzing optical systems for several reasons. Once the system transfer function or point spread function is determined, the output is easily calculated for any input. Also, the system transfer function may be modified to account for additional degradation effects. For the system modeled in this paper, the system PSF is the convolution of the PSF's due to motion, detector area, chromatic aberration, and diffraction.

For circularly symmetric optics and spatially-incoherent imaging, the diffraction PSF is the Airy pattern, proportional to the squared first-order Bessel function of the first kind, also referred to as a squared sombrero function. As per Gaskill[1], the sombrero function is defined as

$$somb(\rho/d) = \frac{2J_1(\pi\rho/d)}{(\pi\rho/d)} \tag{1}$$

where ρ is the radial coordinate in the image plane, measured from the optic axis. An interesting property of the sombrero function is that its volume is the same as the volume of its square:

$$Volume\ [somb(\rho/d)] = Volume\ [somb^2(\rho/d)] = \frac{4d^2}{\pi} \tag{2}$$

All PSF's are normalized with respect to their volumes.

Assume the radiant exitance is centered about the wavelength, λ. (The blurring due to the spectral bandwidth of the radiation is taken into account by the chromatic PSF.) Let D be the diameter of the exit pupil of the imaging optics, and L the distance from the exit pupil to the image plane. The expression for the diffraction PSF is given by

$$PSF_{dif} = \frac{(1/\lambda L)^2 [\frac{\pi D^2}{4} somb(\rho D/\lambda L)]^2}{(\pi D^2/4)} \tag{3}$$

A polar coordinate system is natural for the geometry. The center of the coordinate system is the point on the ground directly below the IR system. Let $\underline{r}$ represent the radial distance from the origin to the center of the detector footprint, and $\underline{s}$ the arc length at a given radius (see Figure 1).

Motion blur is due to the finite integration time of the detector. The widths of the motion PSF account for the area swept out by the footprint during the integration time of the detector. This PSF may be approximated by a two-dimensional rectangle function:

$$PSF_{mot} = \frac{1}{W_s W_r} RECT(s/W_s,\ r/W_r), \tag{4}$$

where

$$RECT(s/W_s,\ r/W_r) = \left\{ \begin{array}{l} 1,\ -W_s/2 < s < W_s/2; \\ \quad\ -W_r/2 < r < W_r/2 \\ \\ 0,\ elsewhere \end{array} \right\}. \tag{5}$$

Let T be the detector integration time, ω the angular velocity, θ the depression angle, and V the fall velocity. The widths for the motion PSF are calculated by:

$$W_s = r\omega T \tag{6a}$$

$$W_r = VT\tan\theta. \tag{6b}$$

The dimensions of the detector will also cause the image to be blurred. The detector integrates over the irradiance of the input scene. The detector of the system described here is square, so the PSF due to the detector dimensions is described by a two-dimensional rectangle function. Since the footprint is centered at the depression angle, θ, the widths of the PSF are different in the radial and tangential directions. The widths of the function are

$$W_{sd} = R(FOV), \tag{7a}$$

$$W_{rd} = R(FOV)/\cos(\theta), \tag{7b}$$

where R is the slant distance from the IR system to the ground at the center of the Field Of View. The corresponding PSF is

$$PSF_{det} = \frac{1}{W_{sd} W_{rd}} RECT(s/W_{sd},\ r/W_{rd}). \tag{8}$$

The last degradation effect to be considered is chromatic aberration. Circular symmetry is assumed, and the blur is approximated by a two-dimensional gaussian function:

$$PSF_{chr} = \frac{1}{R_{chr}^2} Gaus(\rho /R_{chr}), \qquad\qquad (9)$$

where R_{chr} is the $e^{-\pi}$ radius of the chromatic solid angle subtended at the detector, and ρ is the radial distance from the center of the detector. The two-dimensional gaussian is defined here by:

$$Gaus(\rho /b) = \exp[-\pi \ (\rho /b)^2]. \qquad\qquad (10)$$

The overall system PSF is the joint convolution of all the individual PSF's. Each additional PSF included causes the system PSF to be broadened. Even for just two or three PSF's, the system PSF is well-approximated by a normal, or gaussian distribution. This is a property of convolutions described by the Central Limit Theorem[2].

$$PSF_{sys} = PSF_1 \ ** \ PSF_2 \ ** \ ... \ ** \ PSF_n \ --> \ Gaus(\frac{x}{a},\frac{y}{b}), \qquad\qquad (11)$$

where ** implies two-dimensional convolution, yielding a two-dimensional gaussian for the system PSF. The computer algorithm determines the equivalent widths, a and b, by 1) calculating each PSF, 2) digitally convolving all the PSF's together, and 3) finding the points along the x- and y- axis profiles where the amplitude is $e^{-\pi}$ of the central maximum.

Once the system PSF is calculated, it is convolved with the geometrical image irradiance, I", to give the final signal. I" is the blur-free magnified incident irradiance. The convolution progresses in a spiral manner toward the origin specified in Figure 1.

Results and conclusions

Ideally, the output data of the computer simulation would be compared to actual data recorded from an IR system that could be characterized by the input parameters of the algorithm. Although actual data from a scanning IR system was not available, some qualitative analysis shows that the results are as expected.

The computer model is based on the equations given above and was written in Fortran 77 to run on a VAX 11/780. The programs were written for a system which included an image processor, so that it was possible to display both input and output image files. Both actual IR images and computer-generated images were used as inputs to the computer simulation. The grid image shown in Figure 2 proved to be the most useful to analyze the performance of the algorithm. Each white square is 4 x 4 pixels with a gray level of 255 (minimum 0, maximum 255). The system parameters used in the first simulation are listed below:

```
Instantaneous FOV (milliradians) . . . . . . . . . . . . . .      10.0
Effective F#  . . . . . . . . . . . . . . . . . . . . . . .       1.0
Central Wavelength (micrometers) . . . . . . . . . . . . .       10.0
Initial height (meters) . . . . . . . . . . . . . . . . .      100.0
Final height (meters) . . . . . . . . . . . . . . . . . .        1.0
Fall velocity (meters/second) . . . . . . . . . . . . . .        2.0
Angular velocity (revolutions/second) . . . . . . . . . .       40.2
Depression angle from vertical (degrees) . . . . . . . .       30.0
Integration time (milliseconds) . . . . . . . . . . . . .        1.0
Angular width of input image (degrees) . . . . . . . . .       60.0
Recording height of input image (meters) . . . . . . . .      100.0
Samples per scan . . . . . . . . . . . . . . . . . . . . . .    1024
Minimum dimension of input image (pixels) . . . . . . . .       350
Exit pupil diameter (centimeters) . . . . . . . . . . . .        1.0
Detector width (millimeters) . . . . . . . . . . . . . . .        5.0
```

Figure 2. Input Grid

As the detector nears the ground, the blurring effects are reduced because the resolution of the IR system increases due to the decreasing width of the PSF. This is very noticeable in the output image, Figure 3.

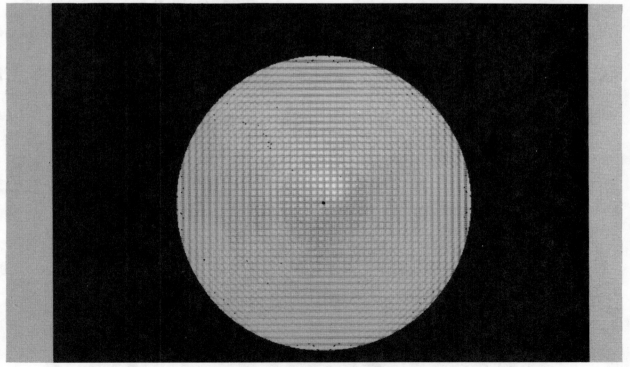

Figure 3. Output Grid Image (IFOV=10mr)

The system input parameters were chosen such that the blurring in the direction of rotation would be dominant over the other blurring effects, a condition that exists in the real data. The number of samples per scan was large, 1024, to enhance the quality of the resulting image. If fewer samples per scan were chosen, "holes" (pixels within the image with zero amplitude or gray scale) in the resulting image would appear. Since the rotational velocity was much greater than the falling velocity, the blurring in the angular direction is much greater than that in the radial direction.

The simulation was run again using the input image of Figure 2 and the same input parameters, except that the Instantaneous Field Of View was increased from 10.0 milliradians to 30.0 milliradians. The result is shown in Figure 4. The blurring in the direction of rotation is still more significant than that in the radial direction because the angular velocity is larger than the fall velocity. However, there is blurring in the radial direction due to the large field of view. The blurring becomes negligible as the footprint approaches the center of the image since the system PSF becomes small compared to the scale of the input image.

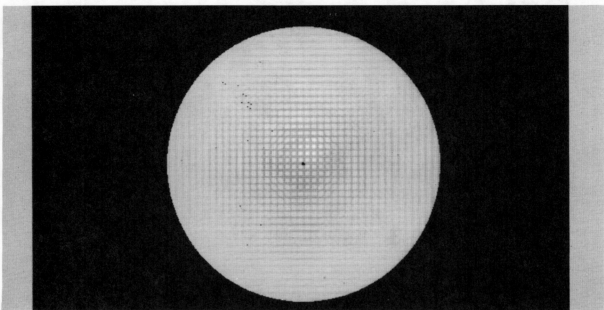

Figure 4. Output Grid Image (IFOV=30mr)

The same two input parameter files were used to produce the results shown in Figures 6 and 7, with the IR image shown in Figure 5 as the input image, an actual IR image. The observations mentioned in the previous paragraph apply for the IR images, again. The increased resolution towards the center of the images can be noticed in the blurring of the two poles at the bottom of the images.

Figure 5. Input IR Scene

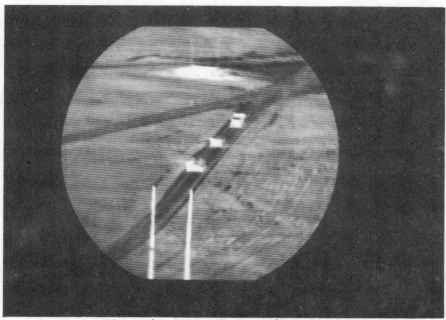

Figure 6. Output IR Image (IFOV=10mr)

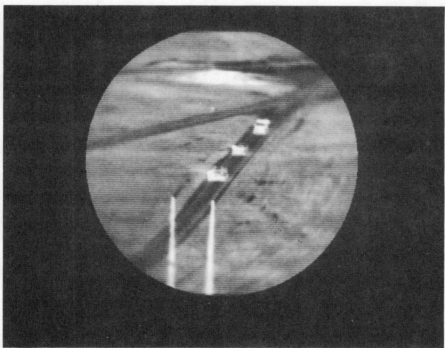

Figure 7. Output IR Image (IFOV=30mr)

The model lends itself to be modified to include other image-degrading effects such as wind shear, atmospheric scattering and turbulence. Each effect may be described by a point spread function and convolved with the other PSF's to yield an equivalent system PSF. There have been very similar Modulation Transfer Functions (MTF's) developed for atmospheric scattering and turbulence [3,4]. (The MTF and the PSF are a Fourier transform pair.)

The goal of this project was to generate, in the manner of a spirally space-variant IR detector, distorted IR data from an "ideal" IR image. This was accomplished, and the programs have been useful in a chain of programs written to study the performance of tracking algorithms.

References

1. Jack D. Gaskill, <u>Linear Systems, Fourier Transforms, and Optics</u> (New York: Wiley, 1978), 398.

2. Athanasios Papoulis, <u>Probability, Random Variable, and Stochastic Processes</u> (New York: McGraw-Hill, 1984), 194-200.

3. Akira Ishimaru, <u>Wave Propagation and Scattering in Random Media</u> (New York: Academic Press, 1978).

4. K. S. Kopeika, "Spatial-Frequency and Wavelength-Dependent Effects of Aerosols on the Atmospheric Modulation Transfer Function," <u>JOSA</u>, Vol. 72, No. 8, August 1982.

Invited Paper

Structure of the Extended Emission in the Infrared Celestial Background

Stephan D. Price

Optical Physics Division, Air Force Geophysics Laboratory
Hanscom AFB, MA 01731-5000

Abstract

The extended emission in the infrared celestial background may be divided into three main components: the zodiacal background, the large discrete sources in the galaxy and the interstellar dust. The zodiacal background is due to the thermal re-radiation of sunlight absorbed by the dust in the solar system. An earth orbiting infrared telescope will detect the diffuse emission from this dust in all directions with maximum intensity lying roughly along the ecliptic plane where the density of dust is highest. Structure with scale lengths of 10° have been measured in both the visual and infrared; finer structure has been detected in the infrared by the Infrared Astronomy Satellite (IRAS). H II regions, areas of ionized gas mixed with and surrounded by dust, are the brightest discrete objects in the galaxy in the long wavelength infrared (LWIR, 7-30μm). The visible radiation from the hot star(s) embedded in these regions is absorbed by the dust and re-emitted in the infrared with a range of temperatures characteristic of the thermal equilibrium for the surroundings of the dust. These regions are relatively large and, if close to the sun, can subtend a significant angular area of sky. The emission from the interstellar dust produces a filimentary structured background, the infrared "cirrus". The observed far infrared color temperature of ~20-35K for the cirrus is consistent with emission from graphite and silicate grains which absorb the interstellar radiantion field. The much larger LWIR color temperature is likely due to a greater abundance of sub-micron particles in the interstellar medium and, perhaps, from band emission due to polycyclic aromatic hydrocarbons. These galactic sources combine along the line of sight to produce an intense band of emission centered on the galactic plane which has full width at half maxima of about 2°.

Introduction

Meter class instruments such as the Infrared Space Observatory (ISO)[1], an approved and funded project of the European Space Agency, and NASA's projected Space Infrared Telescope Facility (SIRTF)[2] are expected to have diffraction limited performance (<20 μr) in the LWIR at high sensitivity. Broad band photometry with these instruments is expected to have a sensitivity of 400 μjy (1 jy = 10^{-26} w $M^{-2}Hz^{-1}$) or about 10^{-21} $wcm^{-2}\mu m^{-1}$ at 11μm for a one second integration time characteristic of area surveys.

The system performance of such LWIR telescopes in near earth orbits will be limited by the general nature and detailed character of the natural background. Trade-offs between the detector size and sensitivity, the amount of data processing and the extent and location of avoidance regions are forced by the intensity and structure of the discrete and diffuse components in the background.

The ultimate sensitivity of any LWIR sensor is set by the photon flux on the detector from the diffuse zodiacal backkground. This emission is pervasive and is observable in all directions for a telescope in near earth orbit. However, the intensity is highly aspect dependent, being brightest near the sun and along the ecliptic plane where the density of the zodiacal dust is greatest. The large source densities and extended structure in the celestial background at the sensitivities projected for space borne infrared telescopes can create a confusion problem which can tax the data processing. The large scale infrared celestial emission, either nearby H II regions or the infrared "cirrus", is wispy in appearance with scale lengths ranging from many degrees down to at least as small as the 0.22mr width of the IRAS detectors. The small scale structure can cause a clutter problem through jitter for a staring instrument such as a mosaic camera or directly for a scanning photometric measurement. The scan clutter problem is further compounded by registration difficulties due to field rotation.

The current observations on the extended celestial background is reviewed with emphasis on the LWIR. Some discussion of the physical processes involved and how they may be modeled is also included.

Zodiacal Emission

The zodiacal background is associated with dust in the solar system. The visual manifestation is due to reflected sunlight and is most prominently seen as an ellipsoidal glow in the twilight of the springtime setting sun or before sunrise in the fall. The infrared background arises from the thermal re-emission of the absorbed sunlight. The geometry of this phenomenon is related to that of the solar system and the pertinent quantities are defined below:

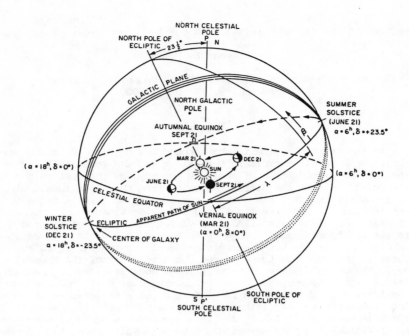

Figure 1.   Co-ordinate Geometry.   β is ecliptic longitude,   λ is ecliptic latitude

ecliptic plane   – the reflex of the earth's orbit on the sky
ecliptic latitude (λ)   – angular measure perpendicular to the plane
ecliptic longitude (β)   – angular measure along the ecliptic plane from the vernal equinox (the
                              intersection of the ecliptic plane and the celestial equator)
elongation (e)   – angular measure from the sun
invariable plane   – the mean plane of the solar system defined as the angular momentum weighted
                              average of all the orbital planes of the planets. The invariable plane plane
                              is inclined 1.25° to the ecliptic with an ascending node (the point where a
                              particle orbiting in the same direction as the planets would cross the ecliptic
                              plane going from negative to positive latitude) of 107° longitude.
astronomical unit (AU)   – the mean distance of the earth from the sun, about 1.5×10^8 km.
albedo (a)   – the ratio of reflected to incident energy of a particle

Visual Zodiacal Light

Until about a decade ago the majority of the data on the zodiacal background consisted of visual observations obtained by ground, balloon borne and satellite based platforms. Comprehensive reviews of these measurement are given by Leinert[3] and Weinberg and Sparrow[4]. Although the absolute intensities obtained by the various experiments show a high degree of internal consistency, they differ from one another by as much as a factor of two at large elongations in the ecliptic plane, a situation which exists even for recent observations. These discrepancies are likely due to systematic errors as Leinert et al[5] found that the visual zodiacal background inside the earth's orbit was constant, to within 2%, during a major portion of a solar cycle.

This constancy also places short term constraints on the dynamics of the dust. Wyatt and Whipple[6] have shown that a typical dust particle will spiral into the sun due to the Poynting-Robertson light drag in about 10^4 years. Constancy in the zodiacal light requires that at least 10^6 grams/sec of dust must be injected into the zodiacal cloud to balance the mass of dust sublimed by the sun or ejected from the solar system[3]. Estimates based on visual observations of the ejection rates of cometary dust and radio measurements on meteors account for only a small percentage of the required dust[7]. However, infrared observations indicate that these sources are more significant that implied by the visual data.

The plane of symmetry defined by the peak visual brightness lies neither in the ecliptic nor the invariable plane. Maucherat et al[8] derived a single plane for the zodiacal cloud at an inclination of 2° from an all sky survey taken by an earth orbiting satellite during a period of over a year. Other measurements [9-11] indicate that within .86 AU of the sun the peak brightness and, consequently, maximum dust density lies close to (i ∼ 2.7°) but not in the orbital plane of Venus (i = 3.4°). Observations by Misconi[11] indicate that outside the orbit of Venus, the plane of symmetry tends to the invariable plane. Thus, the plane of maximum density of the dust in the zodiacal cloud, the symmetry plane, varies in inclination with respect to the ecliptic plane as a function of heliocentric distance.

The local volumetric absorption cross section of the dust can be derived by inverting the brightness integral. If, as is generally assumed, the properties of the dust are independent of position then such an inversion traces the variation in particle density. Usually, the peak intensity of the zodiacal light is assumed to follow a power law ($I \propto R^q$). Leinert et al [12] derived a value of q ~ -2.3 inside the solar circle from observations with the Helios spacecraft. A steeper falloff (q ~ -2.6) was found by Toller and Weinberg[13] between 1 and 2.4 AU from in situ measurements on board Pioneer 10 and 11. Inverting the brightness integral with the above assumptions results in a dust density proportional to $R^{(q-1)}$. In a simple model, dust particles injected into the solar system from a single source acted upon by the Poynting-Robertson effect will have a 1/R distribution[6]. The steeper falloff reported above ($R^{-1.3}$ to $R^{-1.6}$) implies that there are multiple sources distributed over an extended eregion[14] and/or focusing to the ecliptic plane due to collisions[15]. However, Schuemann[16] found that the brightness variation measured by Pioneer 10 was more complex than described by a single power law. Alternatively, Lamy and Perrin[17] adopt a 1/R density distribution and explain the observed brightness variation in terms of an $r^{-0.3}$ variation in the volume scattering function.

Out of the plane of symmetry, the observed visual brightness variations in ecliptic latitude for the region inside the earth's orbit are best represented by a modified fan shaped density distribution[8,17,18].

In addition to the Gegenschein, enhanced back scattering of the dust particles at the anti-solar point, small scale structure has also been observed in the visible zodiacal light by Hong et al[19]. These local variations are reported to be about 10% in amplitude with a 7-10° scale length.

Infrared observations are complementary to the visual in constraining the characteristics of the dust and provide a better probe of the outer regions of the zodiacal clouds. The complex nature of the dust cloud as revealed by infrared observations is reviewed next.

Infrared Measurements

As a dust grain nears the sun its temperature rises until the sublimation point is reached. The grain sublimes, decreasing in size until the radiation pressure balances the Poynting-Robertson drag (d(μm) ρ (g cm^{-3}) ~ 1, Leinert[3]). If the grain temperature is independent of size, a local enhancement will be built up in the region where the grain temperature is near the vaporization point. Several regions of enhanced infrared emission near the sun were reported by ground and balloon based observations in the late 1960's [20-24], the most consistent is at 4 solar radii. A recent multi-color near infrared observations from a balloon platform[25] also detected a ring at 4 solar radii. The relative variation of the near infrared intensities as a function of solar distance are in good agreement between these experiments but the absolute values differ by as much as a factor of four. This discrepancy and the fact that the rings were not detected on all experiments led Koutchmy and Lamy[26] to speculate that the rings may be transitory. At longer wavelengths, several regions of enhanced emission were reported by Lena et al[27] at 10μm near the sun from an experiment flown on the Concorde during the 1973 total solar eclipse.

The near infrared observations of the zodiacal background at larger elongations are limited to two measurements near e ~ 25° obtained on experiments flown in the early 1970's. Hayakawa et al[28,29] obtained multi-color near infrared measurements with a rocket borne experiment while Hofmann et al[30] used a balloon platform for measurements at 2.4μm. More extensive observations are needed on the near infrared zoidacal background as a function of position in order to map the transition from reflected sunlight to thermal emission.

Soifer et al[31] reported the first infrared observations of the thermal emission at larger elongations from the zodiacal dust with 5-6, 12-14 and 16-23μm photometry of a single plane crossing at e ~ 106° from a sounding rocket. The results were rendered somewhat uncertain by the large contribution of off axis earth-shine to the measurement. The same group flew a subsequent spectrophotometric experiment[32] and obtained an 8-14μm spectrum of the emission peak from the plane at e ~ 103°. The spectrum showed a broad silicate emission feature in excess of a 300K gray body continuum. This is consistent with recent laboratory measurements on captured interplanetary dust particles[33] which show prominent 10μm silicate features characteristic of chondritic material.

The largest data base on the LWIR emission from the zodiacal dust cloud is that obtained by the Optical Physics Division of the Air Force Geophysics Laboratory on a series of probe rocket borne experiments from 1974 to 1983 and by the Infrared Astronomy Satellite (IRAS) during 1983. The relative brightness at 11 and 20μm along the ecliptic between 35° and 75° elongation was measured on a September 1974 experiment[34]. These measurements have a systematic zero point error as they are relative to the brightness beyond $|\beta| > 30°$, the region used as a zero reference in deconvolving the ac coupled signals. AFGL obtained LWIR spectrophotometry on several rocket flights. In 1975, several plane crossing were made between 3.5 and 26° elongation[35]. Later, more extensive coverage (22-85° and 137-180° elongation; from -60° ecliptic latitude to the north ecliptic pole) was obtained on the Zodiacal Infrared Project (ZIP)[36], a two flight experiment which sampled the zodiacal background in 15 spectral bands spanning the region from 2 to 30μm. A single plane crossing 19.5° from the sun was measured with a different set of filters during an October 1983 experiment.

IRAS observed the zodiacal emission between 60 and 120° elongation continuously over a 10 month period in 1983[37] in four broad spectral bands, 8-15, 16-30, 45-75 and 85-115μm. The survey aspect was restricted to within 30° of quadrature (the large majority of data lie in the region 75°<e<105°) and the in plane brightness varied by only a factor of 4. On the other hand, the long period of observations with multiple rescans of a given area covered all ecliptic longitudes with a range of viewing geometries.

Figure 2 compares the various brightnesses measured near 11μm along the ecliptic plane. The IRAS data (circles) include the color correction of Hauser and Houck[38]. Spectral photometry at e ~ 22 and 90° from various experiments is compared in figure 3. In general, the various measurements agree to within the quoted errors although there may be a trend for the IRAS in plane observations to be about 40% higher than the ZIP data. The agreement is much better at the pole as can be seen in figure 4. The AFGL experiments measured full width at half maxima (FWHM) of 7, 10 and 70° at e ~ 3.5, 22 and 90°, respectively, which are nearly independent of wavelength over the spectral range of 6 to 25μm. In contrast Hauser et al[37] report IRAS derived values at e ~ 90° of 83° at 12μm and 68° at 25μm.

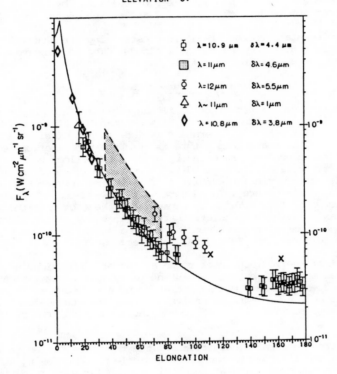

ECLIPTIC RADIANCE
AS A FUNCTION OF
ELONGATION
ELONGATION 0. TO 180.
ELEVATION 0.

Figure 2. In Plane Variation of Zodiacal Brightness at 11μm. The squares with error bars are the ZIP broad band 11μm data[36], crosses are from Soifer et al[31] and Briotta et al[32], and the shaded area are the data of Price et al[34] corrected to the ZIP observations at $|\beta| > 40°$. The circles represent the IRAS observations[37] with the corrections of Hauser and Houck[38]. The near sun measurements of Murdock[35] are shown by the diamonds and the unpublished 1983 rocket borne measurement at e ~ 22° is the triangle with the estimated error. The reference line is the simple model described by Murdock and Price[36] with a 1/R density distribution.

The reference curves in figures 2-4 are from the simple model formulated by Price and Murdock[35] which adopts a mean particle size, albedo and a fan shaped density distribution with a 1/R in plane variation and provides a reasonable fit to the observations. A more rigorous model by Frazier et al[39] assumes a 1/R density variation, a three regime power law distribution of particle sizes and albedos derived from the optical constants of the adopted chemical composition of the material and Mie scattering theory. Although there is reasonable agreement between this model and the in plane visual intensity and the polarization, the infrared values are discordant by at least a factor of two at the anti solar point. Thus the infrared properties of the dust are not simple functions of solar distance[35]. By inverting the brightness integral for the ZIP observations, Hong and Um[40] deduce that, indeed, the density of the dust varies markedly with solar distance. They further concluded that the infrared properties of the dust are not spatially homogeneous and that the visual observations sample a different mixture of dust properties than the infrared. Of course, such conclusions depend on the model assumptions. For instance, Dumont and Levasseur-Regourd[41] interpret the variation of dust properties as due to a change in albedo. They derived values of a = .08 at .98 AU and .06 at 1.4 AU from IRAS measurements and propose that the steeper $R^{-2.3}$ falloff found in the visual represents a shallower density variation (1/R in agreement with the infrared data) coupled with a albedo which decreases with heliocentric distance.

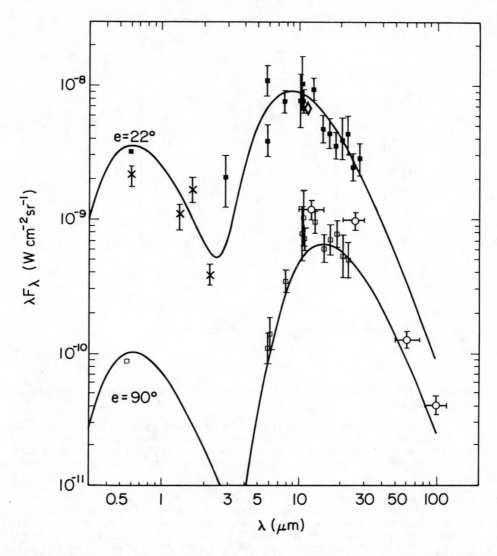

Figure 3.   In Plane Spectra of Zodiacal Emission. Squares, circles and the diamond have the same designations as in Figure 1 (the visual points are taken from Allen's "Astrophysical Quantities"). The crosses are observations analyzed by Nishimura[29].

The plane of symmetry of the zodiacal dust cloud defined by the peak infrared emission lies neither in the ecliptic nor invariable planes, in qualitative agreement with visual measurements. The recent results based on IRAS data for the geometry of the cloud as reviewed by Hauser et al[42] have an inclination of 2-2.5° and an ascending node of 74-78° for the plane near 1 AU from the seasonal variation of the brightness at the ecliptic poles and a 1.5° inclination with a 55° ascending node for the material outside the orbit of the earth based on the longitudinal dependence of the latitude of the peak emission. From an analysis of the IRAS 25μm observations of the peak brightness and the latitude variation with longitude, Dermott et al[43] concluded that the plane of symmetry is either warped or that the infrared zodiacal dust cloud is not axisymmetric. The peak 25μm brightness was well fit by a parabolic variation in elongation.

Also on the global scale, IRAS detected[45] bands of enhanced zodiacal emission centered on the plane of symmetry and 10° either side of it. The 2.2-2.3 AU solar distance of the bands was inferred by their 165-200K color temperature was subsequently supported by photometric parallax measurements of the separation of the bands as a function of elongation[46]. Dermott et al[43,47] have argued that these bands are the geometric projections of dust and small particles associated with the prominent Hirayama asteroid families, Eos and Themis. Sykes and Greenberg[48] calculated that a single collision several million years ago between two members of these families some 15-25km in diameter could produce the band as seen today. Further, given a reasonable asteroidal size distribution and collision rate in the Hirayama families as many as 20 more bands may be detectable in the IRAS data; the more recent events would form arcs rather than complete

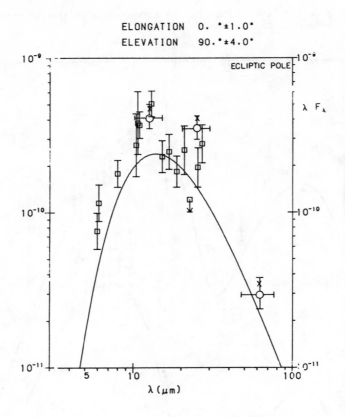

ELONGATION 0.°±1.0°
ELEVATION 90.°±4.0°

ECLIPTIC POLE

$\lambda \ F_\lambda$

$\lambda \ (\mu m)$

Figure 4. Zodiacal Spectrum at the Ecliptic Pole. Comparison of the ZIP (squares) and IRAS (circles) observations at the north ecliptic pole.

bands. The small debris from such an event would collide with the background zodiacal dust resulting in further fragmentation. Collision fragments are removed from the bands to become part of the general background of zodiacal dust when the Poynting-Robertson drag on the resulting particles exceeds about 10 times the gravitational force. Sykes and Greenberg[48] conclude that the majority of the zodiacal dust is produced in this manner.

The IRAS Asteroid Data Analysis System is analyzing several million sources with solar system colors which were detected on a single orbit but not confirmed on a subsequent scan (that is, a potential moving object). Over 10,000 of these observations have been associated with numbered asteroids[49]. Unassociated, high quality observations (in two or more spectral bands with proper color temperature) in this file show a marked concentration to the ecliptic plane[50] which argues that the file probable contains numerous measurements of small asteroids or comets. Dust strewn along the orbit of short period comets should also be an important contributor to this file. For instance, the dust trail from P/Temple 2 was found in the IRAS data by Davies et al[51] as 50 faint seconds confirmed sources detected in July of the IRAS emission and subsequently confirmed on rescanning.

A study of the large scale IRAS sky flux maps by Sykes et al[52] turned up numerous dust trails and streaks in the zodiacal background. In addition to P/Temple 2, dust trains from P/Enke, P/Gunn and P/Schwassmann-Wachmann I have been positively identified. P/Temple 2 has the brightest trail and was detected over at least 48° of the sky, which corresponds to over 1 AU in length. Sykes[53] resolved the comet and tail in an IRAS coadded map and was able to distinguish the dust trail from the large particle component of the tail. Thus, cometary ejecta have a bimodal distribution; a small component sensitive to radiation pressure and a millimeter/centimeter component which remains in the orbit of the comet. The amount of material distributed along the detectable length of the P/Temple 2 trail is estimated at about 10^{12} grams[52] to over 10^{10} kg[53]. This dust was deposited over a period of several hundreds of years by numerous emission events. This is the same order of magnitude as the 5,000 ton mass ejection per perihelion passage estimated by Singer et al[55] for the Enke stream. Sykes et al also found evidence of at least 100 other trails in the sky flux maps but have been able to only tentatively associate trails with P/Temple 1, P/Kopff and P/Shoemaker 2. These trails are generally narrow, only a few of the 2' pixels in the sky flux maps, and most are less than a few degrees in length. The trails vary in brightness, length as ecliptic position. The most distant comet with a dust trail is P/Scwassmann-Wachmann I which was observed at 6.3 AU from the sun.

At 100 to 1000 times the sensitivity of the IRAS survey, future space borne telescopes will have to observe against a structured zodiacal background. The bands, arcs, dust trails, asteroids and comets which populate the IRAS data base will be prominent LWIR objects at these intensity levels.

Extended Galactic Background

The extended LWIR galactic background is primarily due to emission from dust. Galactic gas and dust are concentrated to the galactic plane with a thickness of about 200 pc (pc = 202625 AU = 3.09×10^{18} cm) inside the solar circle and increasing height with galactocentric distance beyond the sun. About 3×10^9 solar masses ($M_o = 2 \times 10^{33}$ grams) of the gas is neutral atomic hydrogen (HI) which is more or less evenly divided between cloud and intercloud material[56]. An equal, or slightly greater, mass of gas is molecular, predominantly molecular hydrogen. Dust comprises only about 1% of the mass of the interstellar material, which itself is less than 10% of the total mass of the galaxy.

Most of the molecular gas is contained to the 6000-plus giant molecular clouds (GMC's) which lie along the spiral arms[57]; the rest is in the numerous smaller clouds distributed throughout the galaxy. The GMC's are massive, $5 \times 10^4 - 10^6$ M_o, gravitational bound objects with an average diameter of 40 pc[58]. About three quarters of the clouds larger than 20 pc in diameter have low kinetic temperature(~ 10K) characteristic of an optically thick cloud in the interstellar environment. The remaining clouds have internal sources of heating due to recently formed stars and are closely associated with H II regions[59]. These molecular clouds generally exhibit a considerable degree of internal structure with extensive fragmentation, numerous condensations and filimentary structures.

H II regions are created around a newly formed O or B star (T_e > 15,000K) or cluster of stars. The stellar uv flux completely ionizes the surrounding hydrogen (hence the designation H II, I represents the neutral ion - II singly ionized, etc.), helium is singly ionized and most of the other elements are either singly or doubly ionized. The ionized electrons are collisionally thermalized and the nebula is cooled and the kinetic temperature is stabilized to about 10,000K by forbidden line emission from the recombination of electrons and ions. The line intensities are large and the H II region is visible over considerably distance. Indeed, these regions are used to estimate extra-galactic distances.

H II regions are prodigious infrared emitters. The majority of the visual and uv flux from the embedded star(s) is absorbed by the dust surrounding the H II region and then is thermally re-emitted in the infrared. The flux from a typical H II region generally peaks at about 70μm but the spectral energy distribution is much broader than blackbody emission at a single temperature with significant LWIR emission. Harris and Rowan-Robinson[60] estimated the core brightness of a typical large H II region at 10^{10} w cm^{-2} μm^{-1} at 10 and 20 μm.

The large size and infrared luminosity of these regions make them easily detectable throughout the galaxy. The closer regions are easily resolved at moderate resolution (< 1mr) and can cover a large area of sky; the Orion complex spans over 100 square degrees in area.

The most pervasive large scale background is the infrared cirrus. This highly structured emission is detectable over the entire sky and is indistinguishable in appearance from the filaments in the outer regions of H II regions and molecular clouds. Gautier[61] described the appearance of the cirrus as having "long, spider-like filaments, clumps and long arching structures composed of small whisps, filaments and clumps. Small scale structure is seen down to the resolution of the IRAS data, 2 to 4 arc minutes". The complex strucutre of the infrared cirrus, molecular clouds and H II regions can present a aspect dependent clutter problem for an area survey with a meter class telescope which can either increase noise or result in spurious sources. The magnitude of the problem depends on the intensity of the filimentary structure at the resolution of the instrument. Armstrong et al[62] determined that electron density fluctuations in the interstellar medium has a density spectrum of the form (size) 3.6 ± 0.2, consistent with a Kolmogorove spectrum (exponent = 11/3) for the energy cascade to smaller sizes due to turbulence. Such a steep function would predict little energy in microfilaments at the resolution of a meter class infrared telescope. The recent measurements on the turbulent spectra of molecular clouds[63] and H II regions[64] indicate a shallower power function than due to Kolmogorove processes. These observations indicate that the interstellar medium is compressible or that there are additional sources of energy as smaller scales.

Gautier[61] noted the correlation between the infrared cirrus and the high latitude reflection nebula studied by Sandage[65] which indicates that some significant fraction of the cirrus is near the sun (~ 100pc) and that the dust re-emits the absorbed integrated star light from the galactic plane. A more quantitative study by deVries and LePoole[66] found that the 100μm cirrus emission correlated well with visual extinction (the amount of dust) derived from star counts for two high latitude reflection nebula and that the infrared temperature of the clouds were extremely constant indicating that starlight, rather than an internal source powered the infrared emission. Sandage's description of the two faint reflection nebula he studied is quite similar to that of Gautier for the infrared cirrus. Sandage noted that one cloud was finely filamented on a scale as small as 30" which indicates that significant cirrus emission is likely to be present on comparable scales.

Correlation has also been found between the infrared cirrus, HI emission[67] and molecular clouds (CO emission)[68]. This was not too suprising since it has been known for some time that interstellar dust, as detected through visual extinction of starlight, is well correlated with the column density of gas (HI plus molecules)[69]. The 20-30K color tempeartures of the infrared cirrus derived from the 60 to 100 μm IRAS observations is consistent with heating by the interstellar radiation field; although Harwit et al[70] have argued that a significant portion of the cirrus emission in these bands are due to forbidden line emission from [O I] and [O III]. The large amount of LWIR emission (IRAS 12 and 25μm bands) [67,71] was however, unexpected. Not only is the LWIR emission orders of magnitude larger than expected on the basis of the long wavelength observations but the 12μm:25μm color temperatures are also high(> 300K).

Part of this emission is probably due to the non-equilibrium heating of small dust grains by the uv in the interstellar radiation field. Draine and Anderson[72] calculated that substantial amounts of continuum radiation at less than 50μm can result in temperature fluctuations in small interstellar graphite and silicate grains. These temperature fluctuations are large when the energy of the absorbed photon is comparable, or greater, than the average heat content of the grain. For the interstellar radiation field this will occur for grains with radii < .01μm. This was found to account for some, but not all, the LWIR excess on a galactic scale[73].

Puget et al[74] suggest that a significant fraction of the interstellar LWIR radiation is due to band emission from polycyclic aromatic hydrocarbons (PAH's). Leger and Puget[75] noted the similarity in the laboratory spectrum of the PAH molecule Coronene and the unidentified emission features at 3.3, 3.4, 6.2, 7.7, 8.6 and 11.3 μm in reflection nebula and some H II regions. PAH's have not yet been observed in the cool interstellar medium although Leger and d'Hendecourt[76] and Crawford et al[77] find a reasonable match between the absorption spectrum of singly ionized PAH's and the diffuse interstellar bands in the visual. On the basis of this investigation Leger and d'Hendecourt conclude that only a small number PAH's are stable in the interstellar medium but that they are the most abundant molecules after H_2 and CO. Duley and Williams[78] on the other hand, claim that PAH's exist only in regions of shock formation since chemical reactions with H and O will effectively destroy them in the interstellar clouds. Instead, hydrogenated amorphous carbon dust will form in denser diffuse clouds and dark clouds. This could, in part, account for the variation of the 12 to 100μm intensity ratio from cloud to cloud and different regions of infrared cirrus.

The overlapping of the emission from the infrared cirrus, discrete and extended sources along the line of sight and in the sensor field of view results in an intense diffuse LWIR emission centered on the galactic plane, a portion of which is shown in figure 5. The emission is nearly constant out to 60° of either side of the center with a nearly exponenetial decrease at larger longitudes. The 2.4 and 4μm background[79] can be accounted for by stellar sources reddened by interstellar extinction. The emission in the far infrared can be attributed to dust in the galactic plane heated either by the interstellar radiation field or hot O and B stars[73]. The LWIR emission along the galactic ridge is again in excess of predictions. If the observed 10μm emission is accounted for by adding circumstellar dust shell sources to the point source model that adequately describes the near infrared background, the ridge brightnesses at 20 and 27μm are overpredicted. If, as suggested by Cox et al[73], a significant portion of the LWIR emission from the galactic plane is from PAH's then their characteristic band features may be observable. Spectrophotometry of the diffuse emission from the galactic plane at a longitude of 36° in figure 6 along with the star burst galaxy M82, for comparison, which has strong PAH band emission. The 375K grey body spectral distribution is characteristic of the 4μm:11μm:20μm color temperature of the emission[80] is shown. The spatial (5'x15') and spectral (1-4μm) resolutions of the measurement is too course for unambiguous interpretation. However, the spectrum is consistent with emission bands from dust at 7.7+8.6μm and 11.3μm in addition to a silicate absorption.

Conclusion

Extended celestial emission will present a complex and varied background for a spaced based, cryogenic LWIR telescope system. Dust arcs, comet trails and asteroids in the solar system provide (moving) sources of spurious signals for area surveys. The highly structured and filigreed background due to structure in molecular clouds, H II regions and infrared cirrus could be a source of scan noise and spurious signals due to field rotation and scan aspect. Evidence indicates that this structure is present down to the resolutions of a meter class instrument and thus cannot be avoided with small fields of view. Strong arguements exists, includign a low resolution spectrum of the diffuse emission from the galactic plane, that emission from this background has a great deal of spectral structure.

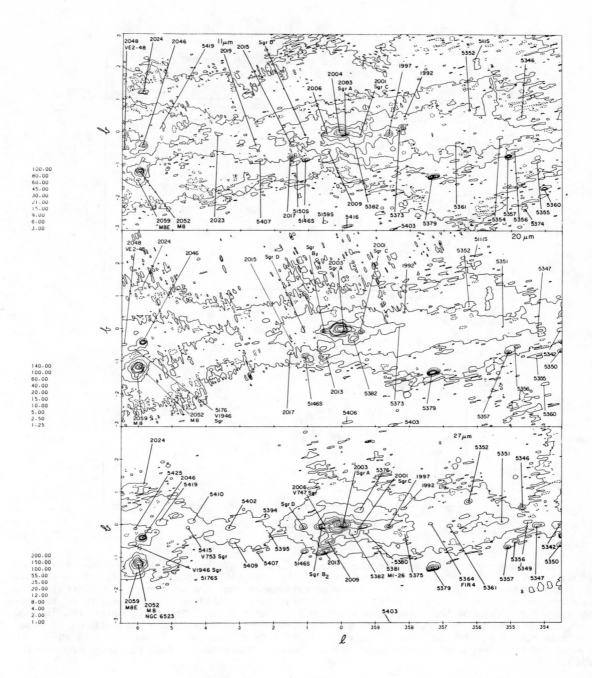

Figure 5. The Diffuse Emission Near the Galactic Center. The 11, 20 and 27μm maps of the diffuse emission near the galactic center. The discrete sources are identified, the unlabeled numbers refer to the AFGL catalog. Contour levels are in units of 10^{-11} w cm^{-2} μm^{-1} sr^{-1}.

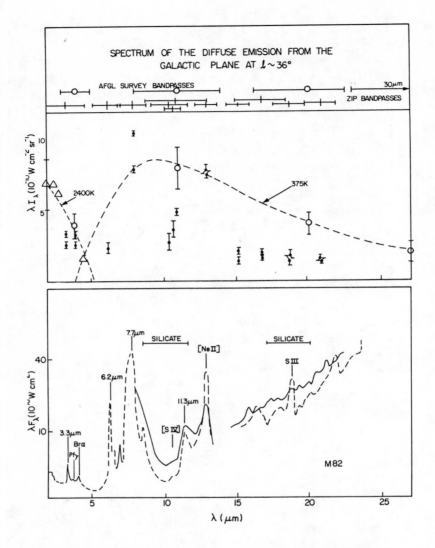

Figure 6. Spectrophotometry of the Galactic Plane. Triangles are balloon[80] and rocket borne[81] near infrared measurements, the open circles are from the AFGL survey[80] and squares the ZIP observations. Error bars are estimated internal errors. ZIP observations at the same wavelength are from separate detectors and are shown individually since the detectors track slightly different regions of the plane. The respective bandwidths of the observations are depicted at the top. The dusty galaxy M82 is shown for comparison of the spectral features. IRAS low resolution spectral data is the solid line at > 8μm. The dashed line is higher resolution data of the central 28" from Willner et al[82] and Houck et al[83] which have been scaled by a factor of two to match the IRAS data. The emission features in M82 have been identified.

References

1. "Infrared Space Observatory", ESA Report Phase A Study, Nov. 1982, (page) t SCI (82) 6, 1982.

2. Rieke, G.H., M.W. Werner, R.I. Thompson, E.E. Becklin, W.F. Hoffmann, J.R. Houck, F.J. Low, W.A. Stein and F.C. Witteborn, "Infrared Astronomy after IRAS", Science, Vol. 231, 807, 1986.

3. Leinert, C., "Zodiacal Light - A Measure of the Interplanetary Environment", Space Sci. Rev., Vol 18, 281, 1975.

4. Weinberg, J.L. and J.G. Sparrow, "Zodiacal Light as an Indicator of the Interplanetary Dust", Cosmic Dust, ed J.A.M. McDonnell, Wiley and Sons, 55, 1975.

5. Leinert, C., I. Richter and B. Planck, "Stability of Zodiacal Light from Minimum to Maximum of the Solar Cycle (1974-1982)", Aston. Astrophys., Vol. 110, 111, 1982.

6. Wyatt Jr., S.P., and F.L. Whipple, "The Poynting-Robertson Effect on Meteor Orbits", Astrophys. J., Vol. 111, 134, 1950.

7. Kresak, L., "Source of Interplanetary Dust", IAU Symp. 90: Solid Particles in the Solar System, D. Reidel, 211, 1980.

8. Maucherat, A., A. Llebaria, J.C. Gonin., "Zodiacal Light, Gegenschein and Sky Background", IAU Colloq. 85: Properties and Interactions of Interplanetary Dust, D. Reidel, 27, 1985.

9. Leinert, C., M. Hanner, I. Richter and E. Pitz, "The Plane of Symmetry of the Interplanetary Dust", Astron. Astrophys., Vol. 82, 328, 1980.

10. Nikolsky, G., S. Koutchmy, P.L. Lamy and I.A. Nesmjanovich, "Photographic Observations of the Inner Zodiacal Light Aboard Salyout 7", IAU Colloq. 85: Properties and Interactions of Interplanetary Dust, D. Reidel, 7, 1985.

11. Misconi, N.Y., "The Symmetry Plane of the Zodiacal Cloud near 1 AU", IAU Symp. 90: Solid Particles in the Solar System, 49, 1980.

12. Leinert, C., I. Richter, E. Pitz and B. Planck, "The Zodiacal Light from 1.0 to .3 AU as Observed by the Helios Space Probes", Astron. Astrophys., Vol. 103, 177, 1981.

13. Toller, G.N. and J.L. Weinberg, "The Change in Near-Ecliptic Zodiacal Light with Heliocentric Distance", IAU Colloq. 85: Properties and Interactions of Interplanetary Dust, D. Reidel, 21, 1985.

14. Leinert, C., "Dynamics and Spatial Distribution of Interplanetary Dust", IAU Collq. 85: Properties and Interactions of Inteplanetary Dust, D. Reidel, 369, 1985.

15. Trulsen, J. and A. Wiken, "Poynting-Robertson Effect and Collisions in the Interplanetary Dust Cloud", IAU Symposium 90: Solid Particles in the Solar System, D. Reidel, 299, 1980.

16. Schuerman, D.W., "Evidence that the Properties of Interplanetary Dust Beyond 1 AU are Not Homogeneous", IAU Symp. 90: Solid Particles in the Solar System, D. Reidel, 71, 1980.

17. Lamy, P.L. and J.-M. Perrin, "Volume scattering function and space distribution of the interplanetary dust cloud", Astron. Astrophys., Vol. 163, 269, 1986.

18. Giese, R.H., G. Kinateder, B. Kneissel and U. Rittich, "Optical Models of the Three Dimensional Distribution of Interplanetary Dust", IAU Colloq. 85: Properties and Interactions of Interplanetary Dust, D. Reidel, 255, 1985.

19. Hong, S.S., N.Y. Misconi, M.H.H. van Dyke, J.L. Weinberg and G.N. Tollner, "A Search for Small Scale Structure in the Zodiacal Light, IAU Colloq. 85: Properties and Interactions of Interplanetary Dust, D. Reidel, 33, 1985.

20. Peterson, A.W., "Experimental Detection of Thermal Radiation from Interplanetary Dust", Astrophys. J. (Letters), Vol. 148, L37, 1967.

21. Peterson, A.W., "The Coronal Brightness at 2.23 Microns", Astrophys. J., Vol. 155, 1009, 1969.

22. Peterson, A.W., "A Determination of the Vaporization Temperature of Circumsolar Dust at $4R_O$", Bull. Amer. Ast. Soc., Vol. 3, 500, 1971.

23. MacQueen, R.M., "Infrared Observations of the Outer Solar Corona", Astrophys. J., Vol. 154, 1059, 1968.

24. Mankin, W.G., R.M. MacQueen and R.H. Lee, "The Coronal Radiance in the Intermediate Infrared", Astron. Astrophys., Vol 31, 17, 1974.

25. Maihara, T., K. Mizutani, N. Hiromoto, H. Takami and H. Hasegawa, "A Balloon Observation of the Thermal Radiation from the Circumsolar Dust Cloud in the 1983 Total Eclipse, IAU Colloq. 85: Properties and Interactions of Interplanetary Dust, D. Reidel, 55, 1985.

26. Koutchmy, S., and P.L. Lamy, "The F-Corona and Circum-Solar Dust Evidences and Properties", IAU Colloq. 85: Properties and Interactions of Interplanetary Dust, D. Reidel, 63, 1985.

27. Lena P., Y. Viala, D. Hall and A. Soufflat, "The Thermal Emission of the Dust Coronal During the Eclipse of June 30, 1973 II. Photometric and Spectral Observations", Astron. Astrophys., Vol. 37, 81, 1974.

28. Hayakawa, S., T. Matsumoto and T. Nishimura, "Infrared Observations of the Zodiacal Light", Space Research Vol X, 249, 1970.

29. Nishimura, T., "Infrared Spectrum of Zodiacal Light", Publ. Astron Soc. Jap., Vol. 25, 375, 1973.

30. Hofmann, W.D., D. Lemke, C.D. Thum and U. Fahrback, "Observations of the Zodiacal Light at 2.4 μm", Nature Phys. Sci., Vol. 243, 140, 1973.

31. Soifer, B.T., J.R. Houck and M. Harwit, "Rocket Infrared Observations of the Interplanetary Dust", Astrophys. J. (Letters), Vol. 168, L73, 1971.

32. Briotta, D.A., J.L. Pipher and J.R. Houck, Rocket Infrared Spectroscopy of the Zodiacal Dust Cloud, AFGL-TR-76-0236 (AD A034 054); D.A. Briotta, Ph.D. Thesis, Cornell University.

33. Sanford, S.A. and R.M. Walker, "Laboratory Infrared Transmission Spectra of Individual Interplanetary Dust Particles from 2.5 to 25 Mircon", Astrophys. J., Vol. 291, 838, 1985.

34. Price, S.D., T.L. Murdock and L.P. Marcotte, "Infrared Observations of the Zodiacal Dust Cloud", Astron. J., Vol. 85, 1980.

35. Murdock, T.L., Infrared Emission from the Interplanetary Dust Cloud at Small Elongations, AFGL-TR-77-0280 (ADC 013 735), 1977.

36. Murdock, T.L., and S.D. Price, "Infrared Measurements of Zodiacal Light", Astron. J., Vol. 90, 375, 1985.

37. Hauser, M.G., F.C. Gillett, F.J. Low, N.T. Gautier, C.A. Beichman, G. Neugebauer, H.H. Aumann, N. Boggess, J.P. Emerson, J.R. Houck, B.T. Soifer and R.G. Walker, "IRAS Observations of the Diffuse Infrared Background", Astrophys. J. (Letters), Vol. 278, L15, 1984.

38. Hauser, M.G. and J.R. Houck, "The Zodiacal Background in the IRAS Data", Light on Dark Matter, ed. F.P. Israel, D. Reidel, 39, 1986.

39. Frazier, E.N., D.J. Boucher and G.F. Mueller, A Self-Consistent Model of the Zodiacal Light Radiance, Aerospace Report TOR-0086 (6432-01)-1.

40. Hong, S.S., and I.K. Um, "Inversion of the Zodiacal Infrared Brightness Integral", Astron. Astrophys., submitted.

41. Dumont, R., and A.C. Levasseur-Regourdd, "Heliocentric Dependences of Zodiacal Emission, Temperature and Albedo", Light on Dark Matter, ed. F.P. Israel, D. Reidel, 45, 1986.

42. Hauser, M.G., T.N. Gautier, J. Good and F.J. Low, "IRAS Observations of the Interplanetary Dust Emission", IAU Colloq. 85: Properties and Interactions of Interplanetary Dust, D. Reidel, 43, 1985.

43. Dermott, S.F., P.D. Nicholson and B. Wolven., "Preliminary Analysis of the IRAS Solar System Dust Data", Asteroids, Comets, Meteors II, ed. C-I. Lagerkqvist and H. Rickman, Uppsala, in press.

44. Deul, E., Physical Modelling of Zodiacal Light, Seminar Notes Kapteyn Lab., 14 Feb, 1986.

45. Low, F.J., D.A. Beintema, T.N. Gautier, F.C. Gillett, C.A. Beichmann, E. Young, H.H. Aumann, N. Boggess, J.P. Emerson, H.J. Habing, M.G. Hauser, J.R. Houck, M. Rowan-Robinson, B.T. Soifer, R.G Walker and P.R. Wesselius, "Infrared Cirrus: New Components of the Extended Infrared Emission, Astrophys. J. (Letters), Vol. 278, L19, 1984.

46. Gautier, T.N., M.G. Hauser and F.J. Low, "Parallax Measurements of the Zodiacal Dust Bands with the IRAS Survey", Bull. Amer. Ast. Soc., Vol. 16, 442, 1984.

47. Dermott, S.F., P.D. Nicholson, J.A. Burns and J.R. Houck, "An Analysis of IRAS' Solar System Dust Bands", IAU Colloq. 85: Properties and Interactions of Interplanetary Dust, D. Reidel, 395, 1985.

48. Sykes, M.V. and R. Greenberg, "The formation of the IRAS Zodiacal Dust Bands", Icarus, Vol. 65, 51, 1986.

49. Tedesco, E., private communication.

50. Rowan-Robinson, M., private communication.

51. Davies, J.K., S.F. Green, B.C. Stewart, A.J. Meadows and H.H. Aumann, "The IRAS fast-moving object search", Nature, Vol. 309, 315, 1984.

52. Sykes, M.V., L.A. Lebofsky, D.M. Hunten and F.J. Low, "The Discovery of Dust Trails in the Orbits of Periodic Comets", Science, Vol. 232, 1115, 1986.

53. Green, S.F., Ph.D. Thesis; quoted by Davies, J.K., "Are the IRAS-detected Apollo asteroids extinct comets?", Mon. Not. Sov. Ast. Soc., Vol. 221, 19p, 1986.

54. Sykes, M.V., private communication.

55. Singer, S.F., J.E. Stanley and P. Kessel, "The LDEF Interplanetary Dust Experiment", IAU Colloq. 85: Properties and Interactions of Interplanetary Dust, D. Reidel, 117, 1985.

56. Baker, P.L., and W.B. Burton, "Investigation of Low-Latitude Hydrogen Emission in Terms of a Two Component Interstellar Gas", Astrophys. J., Vol. 198, 281, 1975.

57. Sanders, D.B., N.Z. Scoville and P.M. Solomon, "Giant Clouds in the Galaxy II. Characteristics of Discrete Features", Astrophys. J., Vol. 289, 373, 1984.

58. Balley, J., "Interstellar Molecular Clouds", Science, Vol. 232, 185, 1986.

59. Solomon, P.M., B.D. Sanders and A.R. Rivolo, "The Massachusetts - Stony Brook Galactic Plane CO Survey: Disk and Spiral Arm Molecular Clouds", Astrophys. J. (Letters), Vol. 292, L19, 1985.

60. Harris, S., and M. Rowan-Robinson, "The Brightest Sources in the AFCRL Survey", Astron. Astrophys., Vol 60 405, 1977.

61. Gautier, T.N., "Observations of Infrared Cirrus", Light on Dark Matter, ed. F.P. Israel, D. Reidel, 49, 1986.

62. Armstrong, J.W., J.M. Cordes and B.J. Rickett, "Density power spectrum in the local interstellar medium", Nature, Vol. 291, 561, 1981.

63. Fleck, Jr., R.C., "A Note on Compressibility and Energy Cascade in Turbulent Molecular Clouds", Astrophys. J. (Letters), Vol. 272, L45, 1983.

64. O'Dell, C.R., "Turbulent Motion in Galactic H II Regions", Astrophys. J., Vol. 304, 767, 1986.

65. Sandage, A., "High-latitude reflection nebulosities illiminated by the galactic plane", Astron. J., Vol. 81, 955, 1976.

66. deVries, C.P. and R.S. LePoole, "Comparison of optical appearance and infrared emission of some high latitude clouds", Astron. Astrophys., Vol. 145, L7, 1985.

67. Boulanger, F., B. Baud and G.D. van Albeda, "Warm dust in the neutral interstellar medium, Astron. Astrophys., Vol. 144, L9, 1985.

68. Weiland, J.L., L. Blitz, E. Dwek, M.G. Hauser, L. Magnani and L.J. Rickard, "Infrared Cirrus and High-Latitude Molecular Clouds", Astrophys. J. (Letters), Vol. 306, L101, 1986.

69. Savage, B.D. and J. Mathis, "Observed Properties of Interstellar Dust, Ann. Rev. Astron. Astrophys., Vol 17, 73, 1979.

70. Harwit, M., J.R. Houck and G.J. Stacey, "Is IRAS cirrus cloud emission largely fine structure radiation", Nature, Vol. 319, 646, 1986.

71. Leene, A., "Warm dust in the R CrA molecular cloud", Astron. Astrophys., Vol. 154, 296, 1986.

72. Draine, B.T. and N. Anderson, "Temperature Fluctuation and Infrared Emission from Interstellar Grains", Astrophys. J., Vol. 292, 494, 1985.

73. Cox, P., E. Krugel and P.G. Mezger, "Principal heating sources of dust in the galactic disk", Astron. Astrophys., Vol. 155, 380, 1986.

74. Puget, J.L., A. Leger and F. Boulanger, "Contribution of large polycyclic aromatic molecules to the infra-red emission of the interstellar medium", Astron. Astrophys., Vol. 142, L19, 1985.

75. Leger, A., and J.L. Puget, "Identification of the "unidentified" IR emission features of interstellar dust", Astron. Astrophys., Vol. 137, L5, 1984.

76. Leger, A., and L. D'Hendecourt, "Are aromatic hydrocarbons the carriers of the diffuse bands in the visible", Astron. Astrophys., Vol. 146, 81, 1985.

77. Crawford, M.K., A.G.G. Tielens and L.A. Allamandola, "Ionized Polycyclic Aromatic Hydrocarbons and the Diffuse Interstellar Bands", Astrophys. J. (Letters), Vol. 293, L45, 1985.

78. Duley, W.W., and D.A. Williams, "PAH Molecules and carbon dust in interstellar cloud", Mon. Not. Roy. Ast. Soc., Vol. 219, 859, 1986.

79. Hayakawa, S., T. Matsumoto, H. Murakami, K. Uyama, J.A. Thomas and T. Yamagami, "Distribution of near infrared sources in the galactic disk", Astron. Astrophys., Vol. 100, 116, 1981.

80. Little, S.J. and S.D. Price, "Infrared Mapping of the Galactic Plane. IV. The Galactic Center", Astron. J., Vol. 90, 1812, 1986.

81. Hayakawa, S., K. Noguchi and K. Uyama, "Near Infrared Multicolor Observations of the Diffuse Galactic Emission", Publ. Ast. Soc. Jap., 1982.

82. Willner, S.P., B.T. Soifer, R.W., Russell, R.R. Joyce and F.C. Gillett, "2 to 8 Micron Spectro-photometry of M82", Astrophys. J. (Letters), Vol. 217, L121, 1977.

83. Houck, J.R., W.J. Forrest and J.M. McCarthy, "Medium-Resolution Spectra of M82 and NGC 1068", Astrophys. J. (Letters), Vol. 242, L65, 1980.

Design and performance of the Halogen Occultation Experiment (HALOE) remote sensor

R. L. Baker, L. E. Mauldin III, J. M. Russell, III

NASA Langley Research Center, Hampton, Virginia 23665

Abstract

HALOE is an optical remote sensor that measures extinction of solar radiation caused by the Earth's atmosphere in eight channels ranging in wavelength from 2.5 to 10.1 micrometers. These measurements, which occur twice each satellite orbit during solar occultation, are inverted to yield vertical distributions of middle atmosphere ozone (O_3), water vapor (H_2O), nitrogen dioxide (NO_2), nitric oxide (NO), hydrogen fluoride (HF), hydrogen chloride (HCl), and methane (CH_4). A channel located in the 2.7 micrometers region is used to infer the tangent point pressure by measuring carbon dioxide (CO_2) absorption. The HALOE instrument consists of a two-axis gimbal system, telescope, spectral discrimination optics and a 12-bit data system. The gimbal system tracks the solar radiometric centroid in the azimuthal plane and tracks the solar limb in the elevation plane placing the instrument's instantaneous field-of-view 4 arcminutes down from the solar top edge. The instrument gathers data for tangent altitudes ranging from 150 km to the Earth's horizon. Prior to an orbital sunset and after an orbital sunrise, the HALOE automatically performs calibration sequences to enhance data interpretation. The instrument is presently being tested at the Langley Research Center in preparation for launch on the Upper Atmosphere Research Satellite near the end of this decade. This paper describes the instrument design, operation, and functional performance.

Introduction

The Upper Atmospheric Research Satellite (UARS) mission is a key part of NASA's middle atmosphere studies program to gather essential atmospheric data that will complement and correlate with data from other spacecraft and supporting sensors. The UARS observatory is a free-flying system that will be launched near the end of this decade and provide data for a period of 18 months. The UARS observatory will be directly inserted by the Space Transportation System into a 600 km circular orbit with a 57 degree inclination. One of the ten experiments on the observatory is the Halogen Occultation Experiment (HALOE). The HALOE instrument is shown in Figure 1. HALOE is designed to monitor globally the vertical distribution of key gases in the ozone chemistry by measuring the extinction of solar radiation in the 2.5 to 10.1 micrometer range passing through the Earth's atmosphere during UARS observatory occultations. Figure 2 depicts this technique. One channel is included to measure absorption in the 2.7 micrometer region for the purpose of inferring tangent altitude pressure. HALOE latitudinal coverage is from 75 deg. S to 75 deg. N, altitude coverage is from 6 to 150 km tangent height, the instantaneous vertical field of view (IFOV) is 2 km in all channels, and the estimated accuracy of retrieved gas mixing ratio is 10 to 15 percent in the mid stratosphere.

The goals of HALOE are to (1) provide fundamental measurements needed to investigate upper atmosphere chemistry, dynamics, and radiative processes, (2) improve scientific understanding of stratospheric ozone depletion due to the ClO_x, NO_x, and HO_x chemical families by collecting and analyzing global data on key chemical species including O_3, HCl, CH_4, H_2O, NO, NO_2, CO_2 (pressure), and (3) study the impact of chlorofluoromethanes (CFM's) on ozone including interactions among the chemical families ClO_x, NO_x, and HO_x by measuring in addition HF. Global measurements of O_3, HCl, CH_4, H_2O, NO, NO_2, and CO_2 (pressure) provide the basis for accomplishing goal 2. The HF measurements, other HALOE results, and synergistic studies with CF_2Cl_2 data from other UARS instruments and the ground correlative measurements program provide the basis for accomplishing goal 3.

The primary scientific mission objectives are to (a) measure and prepare a near-global climatology of the various trace gases previously mentioned, (b) study the global distribution and budgets of the source molecules (CH_4 and H_2O), the reservoir molecules (HCl and HF), O_3, NO and NO_2, (c) determine the response of the upper atmosphere to perturbations due to solar ultraviolet variability, solar proton events, and volcanic eruptions, (d) provide data needed to enhance and refine atmospheric chemical and dynamic models, and (e) distinguish between chlorine produced from all sources and that produced from chlorofluormethanes.

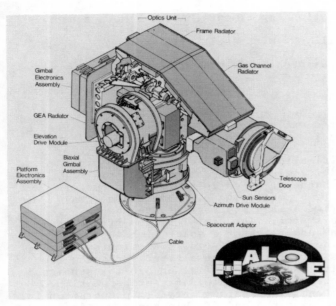

Figure 1. HALOE instrument

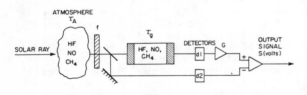

Figure 3. Gas-filter-correlation
radiometry technique

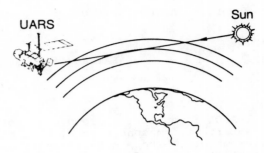

Figure 2. Solar occulation geometry

Table I. Wavelength Characteristics

Gas Filter Radiometer Channels

Specie	Center Wavelength (micrometers)	Bandwidth (cm^{-1})
HF	2.452	117
HCl	3.401	108
CH_4	3.459	82
NO	5.260	62

Broadband Filter Radiometer Channels

Specie	Center Wavelength (micrometers)	Bandwidth (cm^{-1})
CO_2	2.799	128
H_2O	6.616	29
O_3	10.054	78
NO_2	6.250	29

Experiment concept

The HALOE instrument uses the solar occultation experiment technique. As the Sun's rays traverse the Earth's limb during satellite sunrise and sunset events, chemical species in the atmosphere absorb infrared energy in well defined wavelength bands. These tangential absorptions are measured by the HALOE instrument and are later inverted in ground processing to provide gas mixing ratio data.[1]

The satellite-based solar occultation technique has important advantages over ground-based measurements and satellite-based nadir measurements. It provides near global distributions as opposed to point measurements from ground-based in situ techniques. It is self-calibrating since unattenuated sunlight is measured during each event allowing data to be normalized to exoatmospheric values. Also, the longer absorption path in the limb geometry provides higher sensitivity than that from nadir-looking geometry. Moreover, higher vertical resolution is inherent in this technique since most of the absorption occurs at the tangent altitude. Finally, this technique simplifies data interpretation since the Sun is used as a constant background source.

This technique requires precise pointing of the instrument's IFOV. This is accommodated by a 2-axis gimbal system and a sun tracker that can sense a spot on the solar disk without being affected by the combined and varying effects of atmospheric attenuation of solar radiation and changes in the apparent solar extent which occur due to refraction effects. The HALOE pointer/tracker is fully described later.

The HALOE instrument employes two measurement techniques; gas-filter-correlation radiometry and broadband filter radiometry. The species HCl, HF, CH_4, and NO are measured by the gas filter correlation technique while the other gases, O_3, H_2O, NO_2, and CO_2 ,are measured using the broadband filter technique.

The principle of gas-filter-correlation radiometry is illustrated schematically in Figure 3. Solar energy enters the gas correlation section and is divided for each channel into two paths. The first path contains a cell filled with the gas to be measured, and the second is a vacuum path. Each path has a separate detector. An electronic gain adjustment is used in the gas cell path to adjust the difference in detector outputs to zero when there is no target gas in the intervening atmosphere (i.e. above the sensible atmosphere). This condition is called balance and is performed when the instrument's line of sight is outside of the Earth's atmosphere. During occultation when the target gas in the atmosphere is in the instrument's viewing path, a spectral content is introduced into the incoming solar energy which is correlated with the absorption line spectrum of the gas in the gas cell. This correlation upsets the balance causing a difference signal to arise which is proportional to the mixing ratio of the gas in the atmosphere.

The measurement technique requires very narrow wavelength bands centered around the spectral area of interest (species absorption band). The optical bandpass filters specified in Table I are used to isolate these absorption bands. To minimize the sensitivity of the HCl measurement to interferring CH_4 absorption, a CH_4 attenuation cell is placed in the HCl vacuum/gas common path. To provide real-time correction for electro-optical gain shifts or drifts during a measurement, a $1000^{\circ}K$ internal blackbody signal is optically merged with the solar signal and used as an automatic gain control source.

The signal S in volts, neglecting internal sensor thermal emission, is given by

$$S = C \int_{\Delta\phi} \int_{\Delta\theta} \int_{\Delta\nu} N_S A\tau (\tau_A \tau_g \tau_c G - \tau_A \tau_c)\gamma(\theta)d\nu \, d\theta \, d\phi \tag{1}$$

where C is a responsivity factor, N_S is the solar intensity limb darkening curve, A is the aperture area, τ is an optical and electronic attentuation factor including the interference filter dependence, τ_A is the atmospheric transmission, τ_g is the transmission through the blocking gas cell (for the HCl channel only), $\nu(\theta)$ is the spatial instrument function, and $\Delta\nu$ is the spectral bandwidth, and $\Delta\theta$ and $\Delta\phi$ define the angular extent of the (IFOV). The quantity G is the gain adjust parameter used to balance the instrument to zero signal for tangent altitudes above the sensible atmosphere. Equation (1) can be approximated by the expression

$$S = C \overline{N}_S A \overline{\Omega} \, \overline{\tau} \, (\overline{\tau_A \tau_g \tau_c G} - \overline{\tau_A \tau_c})\Delta\nu \tag{2a}$$

or

$$S = C \overline{N}_S A \overline{\Omega} \, \overline{\tau} \, M \, \Delta\nu \tag{2b}$$

where M is the modulation function, Ω is the solid angle of the instrument IFOV, and bars denote averages over $\Delta\phi$, $\Delta\theta$, and $\Delta\nu$.

The conventional broadband radiometer channels are similar to those used for the LRIR and LIMS experiments on the Nimbus 6 and 7 satellites. Each HALOE channel, which uses thermistor bolometer detectors, performs a direct measurement of absorption during the solar occultation event. The bolometer detectors have thermal compensating flakes that correct the measurement for changing ambient conditions.

The measurement equation is the same as equation 2b with M changed to τ_A, the transmission across the atmospheric limb. The ratio of the atmospheric signal to the exoatmospheric signal is a measure of the total mass of the gas in the optical path. Since the exoatmospheric signal is used in the data inversion, the technique becomes self-calibrating for each event.

Exoatmospheric measurements are made for each data event to provide solar limb darkening curve measurements, self-emission measurements, and calibration measurements. These data are necessary to determine with high precision the concentration of gas present in the atmosphere.

The instrument performs the functions depicted in Figure 4 each orbit. First, the pointer/tracker system places the instrument IFOV 4 arcmin below the solar top edge and centered on the solar disk in azimuth. For a sunset event, the gas correlation channels are then electronically balanced. Immediately after balance, five vertical scan cycles of the Sun are performed. These data are used later in ground processing to correct for atmospheric refraction effects and Sun spots. After the scan cycles, a calibration sequence is performed where various gas cells and neutral density filters are inserted into the optical path to provide a measure of scale factor stability for each channel. During deep space looks, data is gathered to correct for self emission, crosstalk, and other instrument effects.

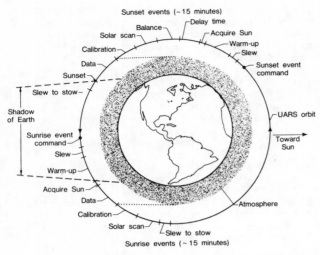

Figure 4. HALOE orbital events sequence

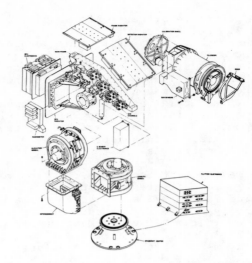

Figure 5. Exploded view of HALOE instrument

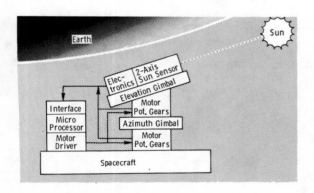

Figure 6. HALOE pointing system
block diagram

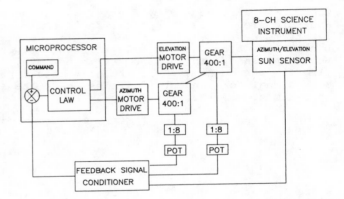

Figure 7. Pointer/tracker control system

Functional and electromechanical description

First, an instrument overview will be given, including a brief description of instrument electronics, operational software, and the pointer/tracker subsystem. Then, each instrument operating mode will be described. Other instrument performance aspects will be presented with the description of instrument operating modes.

An exploded view of the HALOE instrument is given in Figure 5. Overall characteristics are given in Table II. Main instrument subassemblies include the spacecraft adaptor, azimuth gimbal, elevation gimbal, sun sensor, gimbal electronics assembly, platform (off-gimbal) electronics assembly, optics assembly, and telescope. HALOE/UARS command, power, and telemetry interfaces are summarized in Table III. HALOE receives two redundant, regulated 28 V inputs from the spacecraft for both primary and heater power. Internal instrument timing is derived from a 1024 kHz clock and a 0.5 Hz major frame pulse supplied by the spacecraft. Two impulse commands are used for routine orbital operation and sixteen-16 bit serial digital commands are available for instrument initialization and reformatting. These commands are also extensively used for ground test purposes. The data system digitizes the science and engineering data (12 bit quantization) into a NRZ-L coded serial PCM telemetry stream of 4000 bits/second which contains measurements of 127 parameters. Additionally, 22 analog housekeeping measurements and 7 discrete (bi-level) status checks are provided to the spacecraft interface for telemetry.

HALOE instrument functions are provided by two RCA 1802[2] microprocessors, which contain software routines that acquire and track the Sun, acquire and format analog data for telemetry, and sequence the instrument through various modes during the sunrise and sunset events. The microprocessor memory consists of 4096 bytes of PROM (programmable read-only memory) and 3072 bytes of RAM (random access memory). HALOE software is organized into two basic sections, power-on/initialization routines and run-time routines. The power-on/initialization routines initialize the microprocessor, the run-time software, and instrument hardware. The run-time routines provide all control functions for the instrument and contain the command handler which provides external communication via the serial digital commands. A time multiplexed format allows simultaneous control of the pointer/tracker, data acquisition, and internal mode sequencing. An executive program allows provisions for ground testing, calibration, and troubleshooting via the serial digital commands. Also, one of the serial digital commands can be used to manually correct for some instrument malfunctions that may occur in orbit.

Coarse and fine sun sensors provide error signals to the closed-loop microprocessor based control system which independently drives azimuth and elevation gimbals to point the instrument at the Sun. Block diagrams of this system are given in Figures 6 and 7. The stepper motor driven azimuth gimbal, which is mounted directly to the spacecraft adaptor, can be positioned within a ±185 degree azimuth range. The elevation gimbal, which has a design construction nearly identical to the azimuth gimbal, is mounted between the azimuth gimbal and the instrument optical assembly and can be positioned within a 39 degree elevation range. These gimbals are pinned for launch protection and the pins are removed after launch by "one shot" pyrotechnic actuators. Electrical "soft" stops are used to limit the azimuth and elevation angular range to protect flexure cables which route all electrical signals through the gimbals. Mechanical "hard" stops are located just beyond the soft stops to provide protection when instrument power is off.

HALOE is activated twice each orbit--once during satellite sunrise and once during satellite sunset. Normal operation requires a "sunset" impulse command for a sunset event and a "sunrise" impulse command for a sunrise event. These commands initiate an automatic sequence through the operating modes shown in Figure 4. The sunset sequence begins with a slew from the day stow gimbal position to the expected sunrise gimbal position, followed by solar coarse acquisition/track, fine acquisition/track, a programmed time delay, an instrument balance, 5 solar scan cycles, a calibration sequence, a science data retrieval period, and finally is concluded with a slew to the night stow location. The sunrise sequence occurs basically in the reverse order, with three exceptions. Time delay and balance modes do not occur in the sunrise sequence. One additional mode, boresight check, occurs immediately prior to the slew to day stow position for the sunrise sequence only.

In the discussion that follows, all modes in the sunset automatic sequence will be described in the order in which they occur, and any differences in sunrise operation will be noted. The first mode in this sequence is the "slew" mode, which is followed by "coarse acquisition/track". In the "slew" mode, the gimbals are positioned from their day stow positions to the expected sunset location. An actual potentiometer position is compared to the desired position to generate an error signal for the control system. The actual position is sampled by the microprocessor at an 8 Hz rate and 30 pulses are issued to the stepper motor in the desired direction for minimizing this error. Updating at the 8 Hz rate provides a slew rate of 240 steps/sec. Since the motor rotates 1.8 deg/step with the output shaft driving a 400:1 reduction gear, one step equals a gimbal rotation of 16.2 arcsec and the effective slew rate is 1.08 deg/sec. This mode continues until the gimbals are aligned to the expected sunset position within ±3 arcmin in elevation and ±6 arcmin in azimuth. The expected sunset position is initially chosen to place the gimbals within their +3 deg solar acquisition field-of-view. At this point, the sequence enters the "coarse acquisition/track" mode. Solar acquisition occurs immediately in azimuth, but the control system logic does not enable elevation acquisition until the coarse sun sensor error signal changes polarity from "-" to "+". Sunrise elevation acquisition does not wait for this polarity change, and occurs immediately. After coarse acquisition occurs, the actual position is sampled by the microprocessor at an 8 Hz rate, and either 0, 2, or 4 pulses are issued to the stepper motor in the direction to minimize error (control law algorithm). Updating at the 8 Hz rate provides a maximum track rate of 32 steps/sec which is equivalent to a gimbal rate of 8.64 arcmin/sec.

The control system error signals for the "coarse acquisition/track" mode are provided by coarse sun sensors. They consist of a three element silicon photodiode detector arrangement located behind two masks. For descriptive purposes consider the two outer detectors, detector A and B, and the center detector, detector C. The solar rays illuminate one of the outer detectors (A) more than the other outer detector (B) unless

Table II. HALOE Instrument Characteristics

Type of instrument:	Gas filter correlation and broadband filter radiometers
Type of Measurement:	Solar occultation infrared atmospheric absorption
Channels:	8: HF, HCl, CH_4, NO, CO_2, H_2O, O_3, NO_2
Detectors:	6 InAs, $2HgCdTe$, and 4 Thermistor Bolometers
Telescope:	5.8 inch, f/6.6, 37.8 inch FL Cassegrain
Instantaneous Field-of-View:	2.0 arcminutes elevation
	6.0 arcminutes azimuth
Vertical Resolution:	1.2 miles (2 km) at limb
Horizontal Resolution:	3.8 miles (6.3 km) at limb
Pointing Accuracy:	30.0 arcsec elevation
	30.0 arcsec azimuth
Pointing Knowledge:	12.0 arcsec elevation
	15.0 arcsec azimuth
Instrument Weight:	216 pounds (maximum)
Instrument Power:	130 watts (orbit average)
Data Rate:	4000 bps

Table III. HALOE Command, Power, and Telemetry Interfaces

HALOE Inputs	HALOE Outputs
● Primary Power	● 1 Serial Digital Science Data Channel
● Heater Power	● 22 Analog Housekeeping Channels
● Survival Power	● 7 Discrete Status Channels
● 1.024 MHz	
● Major Frame Sync	
● 2 Impulse Commands	
● 1 Serial Digital Command	

the solar beam is aligned with the masks. Thus, the combined detector output, A-B, is used as an error signal in the closed-loop control system. Actually, the A-B signal is divided by the center detector output, C, and this analog error signal is sampled by the microprocessor to provide the digital track error that enters the control law algorithm. Dividing by C normalizes the error to solar intensity which changes during the occultation event, and provides a "sun presence" signal to enable the coarse track mode when the Sun enters the coarse sun sensor field-of-view. This field-of-view is set to ±3 deg by appropriately sizing the front mask.[3] These detectors operate wideband in wavelength over a dynamic intensity range of 1000:1. Two orthogonally mounted coarse sun sensors provide independent error signals to the azimuth and elevation control systems.

After the sunset (or sunrise) automatic sequence enters the "coarse acquisition/track" mode, the control system closes the azimuth and elevation errors until the gimbals align the HALOE instrument optical axis to within 0.3 deg of the solar radiometric centroid. Then, elevation gimbal control is passed from the elevation coarse sun sensor to the elevation fine sun sensor, and the instrument enters the "fine acquisition/track" mode in the automatic sunset sequence. When the instrument enters this mode, the azimuth solar position is stored in a register that becomes the new expected azimuth solar position for the next sunset event. The instrument contains two of these azimuth registers--one for sunset events and one for sunrise events. The azimuth gimbal positions that are stored in these registers easily align the azimuth gimbal within the ±3 deg coarse sun sensor acquisition field-of-view. Also at this moment for sunset events, the microprocessor calculates the expected elevation angle for the next sunset event and stores this value in a third register. A fourth register contains a fixed value for sunrise expected elevation solar position. Expected elevation solar position is constant for sunrise events but changes with beta angle for sunset events.

After the instrument enters the "fine acquisition/track" mode, the elevation gimbal continues to align the HALOE optical axis until the solar top edge is registered, and the azimuth gimbal continues to align the optical axis until it is within +0.5 arcmin of the azimuth solar radiometric centroid. The azimuth and elevation gimbals maintain these alignments through the rest of the automatic sequence (except for the scan mode which will be described later). A separate all-digital sun sensor provides the error signal for the elevation gimbal "fine acquisition/track" mode. An f/12.5, 318 mm focal length Galilean telescope images the solar disk on a linear 256 element silicon photodiode array, which yields a field-of-view of 1.2 deg. An interference filter having a 30 Angstrom spectral bandwidth is used to set the operating wavelength to 700 nm. The dynamic intensity range of this device is 100:1 and is set to 80% of the photodiode saturation level with a neutral density filter located in front of the interference filter. Diode-to-diode spacing in the array is 25 microns so that the exoatmospheric Sun illuminates 118 diode elements. As the solar extent shrinks during a sunset occultation due to atmospheric refraction, the number of diodes illuminated decreases until approximately 25 remain when the instrument IFOV reaches a tangent height of 6 km in the atmosphere. The telescope focal length was sized to provide each diode with a 16.2 arcsec angular spacing which corresponds to one stepper-motor step to the elevation gimbal. The video output from the self-scanned photodiode array is processed through an adaptive threshold electronics signal processor.[4] This signal processor converts the video signal into a top and bottom solar edge address. In the "fine acquisition/track" mode, the actual top edge address is compared to a desired address (which is programmable) to provide an error signal to the control system. Motor steps, which are equal to one diode spacing in the array, are issued from the control law algorithm until the solar top edge is at the desired address. For normal operation, the desired location for the top edge is set to diode 125. This sets the instrument IFOV 4 arcmin below the solar limb since the fine sun sensor diode 140 is boresighted to the instrument optical axis. In the automatic sunrise or sunset instrument operating sequence, elevation control is passed from the coarse sun sensor to the fine sun sensor when the error is closed to within 0.3 deg. In the sunset event only, control is passed back to the coarse sun sensor if the fine sun sensor error goes greater than 8 diodes.

Following the "fine acquisition/track" mode in the automatic sunset sequence are the "time delay" and "balance" modes, neither of which are used in the sunrise sequence. Time delay is simply a programmable time delay between fine acquisition and balance. This delay is updated by ground command to adjust the automatic sequence for long or short occultation events. Occultation events are short at spacecraft beta angles near zero and grow progressively longer as beta angle increases. The balance mode was essentially described in the experiment concept section. Four of the eight radiometric science channels are gas filter correlation radiometers, and the other four are conventional broadband filter radiometers. Each of the gas correlation channels have two optical paths--one which contains a gas cell filled with the target gas, and one which does not. In order to compensate for slowly varying non-spectral long term drifts in instrument performance such as detector responsivity changes or amplifier gain changes, each of the gas correlation channels contain an automatic gain control (AGC) loop (with a 20 second time constant) which is used during a sunset event to "balance" the instrument. This is accomplished by the microprocessor which samples the difference signal between the two paths and applies successive gain corrections with the AGC until the difference is within acceptable limits nominally 1 to 4 times the noise level. Gains are then held constant at the balance value by the AGC circuit during the total orbit. This will be discussed in more detail in the next section.

Following the balance mode in the automatic sunset sequence is the "scan" mode and then the "calibrate" mode. In the "scan" mode, a constant 2.16 arcmin/sec scan rate in both up and down directions is accomplished by offsetting the fine track error signal with a ramp function. This ramp is an 8 bit software counter which is either added to the top solar edge address or the bottom solar edge address to provide the desired scan amplitude. Scan amplitude is set to 40 arcmin to provide exoatmospheric solar limb darkening data in the science channels before sunset events and after sunrise events. A calibrate mode is used to place known amounts of absorption in the science channels to facilitate inversion of the science data. The instrument optical assembly has a calibration wheel that contains twelve apertures; one aperture is open, three contain neutral density filters, and eight contain calibration gas cells. This stepper-motor driven wheel is rotated so that each aperture is successively positioned into the optical path for a 7.5 sec duration. This is accomplished in the automatic sequence while tracking the exoatmospheric Sun which becomes the radiometric calibration source. Data through the apertures are used as a calibration check for the science channels. The drive system is also designed to allow manual stepping of the apertures into the beam one at a time, and to allow the wheel to be returned to the open port with a "cal home" command. With the proper in-orbit initialization, the entire calibration cycle can be omitted if desired.

At the conclusion of the calibrate mode in the automatic sunset sequence which leaves the calibration wheel in the open position, the instrument enters a "data retrieval" mode followed by a slew to the night stow position. Timing of the impulse sunset command, calculation of the expected sunset elevation position, and time delay setting are used to adjust instrument modes before the data retrieval period so that the calibrate mode is just completed as the tangent path reaches 150 km in the atmosphere. Data retrieval begins at 150 km and continues to 6 km or until the coarse sun sensor is blocked by clouds. During data retrieval, the fine sun sensor controlled elevation gimbal keeps the instrument optical system aligned to the top edge of the sun within ±0.5 arcmin (2 diode elements). However, top edge telemetry data can be interpolated to provide pointing knowledge to within 12 arcsec during the event. The azimuth coarse sun sensor controlled azimuth gimbal keeps the instrument IFOV centered with respect to the solar radiometric centroid within ±0.5 arcmin. The elevation IFOV begins the event at 4 arcmin below the top solar edge, and due to the atmospheric refraction effects, gradually moves along the solar diameter toward the bottom edge during the sunset event. Data retrieved during the scan mode is used to correct the science data for solar limb darkening effects while the fine sun sensor solar extent data (top and bottom edge position) is used to register the science data in an altitude grid. Event termination occurs when the coarse sun sensor drops below a programmable "sun presence" intensity threshold. When this threshold is reached, the coarse sun sensor passes control of the gimbal position back to the potentiometer in the closed-loop control system. Then, the gimbals slew to programmable positions used for night stow. Night stow position is carefully chosen to provide thermal and contamination protection between the sunset automatic sequence and the sunrise automatic sequence. Sunrise event termination occurs when a programmable elevation angular threshold is reached while in the data retrieval mode that coincides with a tangent path altitude of 150 km in the atmosphere. Event termination initiates the remainder of the sunrise automatic sequence, consisting of calibrate, scan, boresight check, and slew to day stow position. Since the instrument cannot be rebalanced before the sunrise sequence, instrument circuitry is carefully designed to provide minimum drift between the sunset balance mode and the end of a sunrise event. A day stow position is chosen to provide thermal and contamination protection between the end of a sunrise and the beginning of a sunset. The "boresight check" mode is only performed during sunrise events and is used as a monitor of the azimuth coarse sun sensor boresight alignment to the fine sun sensor's IFOV. In the boresight check mode which occurs immediately prior to the slew to day stow, the azimuth control system error signal is offset by +10 arcmin for 3 seconds and then -10 arcmin for 3 seconds. Azimuth coarse sun sensor and fine sun sensor data are analyzed for the +10, 0, -10 arcmin relative azimuth positions to determine boresight error. This boresight error can be programmed into the azimuth control system by ground command as an offset to provide realignment of the fine sun sensor's IFOV to the solar disk center. A intermittant check of the instrument's IFOV to the fine sun sensor's IFOV is performed in a similar manner. The major difference is that the solar disk is scanned at each azimuth offset angle. Thus, any drift in alignment while in orbit will be detected and corrected when required.

Optical description

The optical assembly is contained on the mainframe as shown in Figure 5. An optical schematic is given in Figure 8. A rectangular field stop at the focal point of a 5.8 inch diameter Cassegrain telescope determines the HALOE rectangular 2x6 arcmin science IFOV. This 96 cm focal length, f/6.6 telescope achieves diffraction limited optical performance at the science channel operating wavelengths. The field stop is reflective to provide rejection of out-of-field solar energy. This feature reduces stray light levels and minimizes thermal loading inside the instrument due to solar radiation. An active thermal control system maintains the aluminum telescope housing and optics mainframe at 21°C to minimize loss of performance due to temperature gradients. Survival heaters are mounted on the back of the silver coated primary and secondary mirrors to prevent these components from cooling below their minimum critical temperature when instrument power is off. Other survival heaters similarly protect other parts of the instrument. The secondary mirror is mounted to a spider and the open aperture is closed off with a door to prevent contamination during launch. A one-shot pyrotechnic actuator is used to open the door after the post-launch spacecraft contamination cloud has dispersed. The telescope barrel contains a series of black, concentric baffles to minimize stray light. One of these baffles, which coincides with the secondary mirror position, serves as the aperture stop for the HALOE optical system. This baffle is imaged on each detector to minimize signal non-linearity due to detector response nonuniformity. The HALOE telescope is rigidly mounted in two planes to a flat, aluminum honeycomb structure referred to as the

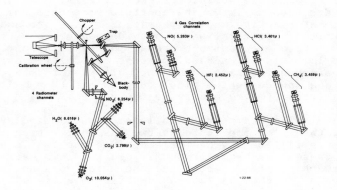

Figure 8. HALOE optical schematic

mainframe. All HALOE optical components are mounted either directly or on subassemblies that are rigidly mounted to the mainframe. Most of the optical elements are mechanically retained on kinematic mounts which are potted after alignment. Optics on the mainframe can be generally divided into four categories: common optics, reference blackbody optics, radiometer channel optics, and gas correlation channel optics.

The common optics consists of two transfer lenses, a calibration wheel, and a chopper. A transfer lens, which is made of germanium as are all of the optical lenses, recollimates light from the field stop and reimages the aperture stop at the calibration wheel location in the optical path. This lens also serves as a blocker to protect optical surfaces and coatings from ultraviolet radiation damage. All lenses have anti-reflection coatings to improve throughput efficiency. The calibration wheel contains one open aperture, three neutral density filters, and eight gas filled cells (representing two gas concentrations for each of the four gas correlation channels). As described earlier, the neutral density filters and gas cells are rotated into the optical path during the calibration mode, to inject known amounts of absorption into each of the HALOE channels. In the science data retrieval mode, the optical beam goes through the open aperture in the calibration wheel. A second transfer lens is used to refocus the field stop at the chopper location. The chopper has a dual track circular blade to chop solar radiation from the telescope on the inner track at 145.8 Hz and radiation from the reference blackbody on the outer track at 291.7 Hz. The chopper surface is polished and gold coated to alternately pass the solar radiation for the gas correlation channels and reflect the solar radiation for the radiometer channels. This metal blade is accurately machined and polished to provide a 50% duty cycle signal for all channels, which also minimizes solar signal crosstalk into the 291.7 Hz demodulated reference blackbody signal. An LED pickoff is used to drive the detector's synchronous demodulators in the signal processing electronics. A synchronous motor rotates the blade continuously at 1250 rpm once the instrument is powered.

A 1000°K reference blackbody is used in the HALOE instrument as an independent measure of gas and vacuum path transmissions for the gas correlation channels. The difference in gas and vacuum path detector signals from the blackbody is held constant once the detector solar signals are balanced. This is accomplished in a closed-loop feedback mode, as described earlier, using the AGC. Maintaining a constant relationship from the known blackbody source assures that the difference in solar signals from the two paths are due to atmospheric gas species and not due to changes within the instrument. The blackbody, which is manufactured at the Langley Research Center, consists of an alumina core that is heated to 1000°K with platinum resistance wire, which is also used as the controlling element in a bridge circuit to maintain a core temperature stability of less than $\pm0.5^\circ$K over a 15 minute period and provide long term drift of less than $\pm12^\circ$K over the 2 year life. An inconel structure supports the core and provides a source aperture of 1 mm diameter. This aperture is imaged on the outer track of the chopper blade. A beamsplitter is then used to fold this reference beam into the optical path and make it colinear with the solar beam. Finally, a beam trap retroreflector is used to provide the radiometer channels with a thermally stable target during their open chopper cycle. This trap also collects solar energy reflected by the beamsplitter and eliminates this potential stray light source.

The four radiometer channels operate as conventional radiometers, using interference filters for spectral separation and thermistor bolometers for detectors. A relay lens is used to recollimate the solar beam for the interference filters, and three element germanium immersion lenses are used to image the aperture stop onto the bolometer active thermistor flakes. The bolometer active and compensating flakes are used in bridge circuits with current and voltage monitors, which provides accurate determination of bolometer responsivities. Bolometer outputs are ac couped to synchronous demodulators which provide high gain and low noise. Each bolometer's demodulated output passes through a 1.16 Hz low pass filter before being sampled by the multiplexer at an 8 Hz rate. These outputs along with the voltage and current monitor outputs are then multiplexed into the science serial digital data stream for telemetry to the ground.

The colinear reference and solar beams are recollimated and routed by mirrors and relay lenses so that each side of the mainframe contains two gas correlation channels. Dichroic beamsplitters and interference filters divide the beam into the four spectral channels given in Table I. Each of these four beams is subdivided into two beams, one which passes through a gas cell before reaching its detector and one which does not. These are referred to as the vacuum path and the gas path as described earlier.

The gas cells have cylindrical housings and flat sapphire windows. These windows are wedged, as are the neutral density filters, interference filters, and beamsplitters to reduce interference effects (channeling) within each channel's spectral response characteristic. A gold housing is used for the HF and HCl gas cells to preserve the gas fill concentrations over a multi-year life. The CH_4 and NO gases use gas cells made of pyrex glass. The HCl channel uses a CH_4 blocking cell in the gas/vacuum common path to reduce spectral interference effects of CH_4 in the HCl bandpass. Each gas cell is carefully filled to the desired gas concentration and then accurately characterized with a fourier transform infrared (FTIR) spectrophotometer.[5] Relay lenses are used to recollimate the beam for the interference filters and gas cells. The NO channel contains a sapphire blank equal to the thickness of the gas cell windows to compensate the vacuum path for spectral response shaping caused by the sapphire cell windows in the gas path. Sapphire does not significantly alter spectral response at the wavelengths of the other channels.

Two element germanium field lenses reimage the aperture stop onto the photovoltaic detectors. Three of the channels use InAs detectors while the fourth channel, NO, uses HgCdTe detectors. The photovoltaic detector elements are mounted to thermo-electric coolers. The closed loop thermal control system provides a $0.6^{\circ}K$ operating temperature range at $203^{\circ}K$ for the InAs detectors and $195^{\circ}K$ for the HgCdTe detectors. A thermistor within the evacuated detector cavity is used as the controlling element in a bridge circuit that adjusts cooler power to maintain the temperature range. These coolers are mounted in close proximity to the large passive radiator that is attached to the mainframe so that the cooler power dissipation does not cause temperature gradients.

Detector outputs are amplified and synchronously demodulated at 145.8 Hz to retrieve the incoming solar signal and at 291.7 Hz to retrieve the reference blackbody signal. Telemetry outputs from each of the four channels are V, R, ΔV, and ΔR where V is the vacuum path detector output from the solar signal, R is the vacuum path detector output from the reference blackbody signal, ΔV is the difference in gas/vacuum detector outputs from the solar signal, and ΔR is the difference in gas/vacuum detector outputs from the reference blackbody signal. The differenced outputs have a factor of 60 more gain (37 for the CH_4 channel) than the vacuum path detector outputs, and in addition, the ΔR signal is maintained constant during the data event after the gas/vacuum detector ΔV signals are balanced by means of an AGC loop that controls the gain of the gas path signals. The V and ΔV signals pass through 1.16 Hz low-pass filters and are telemetered at an 8 Hz rate. The R and ΔR signals pass through 0.23 Hz low pass filters and are telemetered at a 1 Hz rate. Also telemetered for each channel are the AGC signal and a high resolution output of the AGC signal during the event. Other primary telemetry channels used for gas correlation channel science data reduction are the reference blackbody voltage and current, and the detectors' temperature. These outputs are all multiplexed into the serial digital data stream.

Conclusions

The HALOE instrument has been successfully assembled and optically aligned by the Project team at the Langley Research Center. The radiometric performance test series was recently completed and the data are in final evaluation. Initial conclusions are that the instrument is capable of making all required measurements. Currently, the instrument is in the environmental test series. After all testing is complete, the HALOE instrument will begin a refurbishment cycle to replace limited life components (gas cells and detectors), to install precisely characterized bandpass filters, and to set final electronic gains. The instrument will repeat the full radiometric and environmental test series before shipment to General Electric for integration on the UARS observatory.

Acknowledgements

The initial design of this instrument was performed by TRW, Inc., under contract NAS1-15880. Design finalization and hardware development, fabrication, assembly, alignment, and test was accomplished by the HALOE team at NASA Langley, which is managed by Mr. J. L. Raper for the Principal Investigator Dr. J. M. Russell, III. The authors thank all members of the HALOE Project team for their dedication during the development, assembly, and alignment phases and especially for their exceptional efforts during the radiometric testing phases in resolution of problems and demonstration of the capability and character of the instrument.

References

1. Russell J. M., Park J. H. and Drayson S. R., "Global Monitoring of Stratospheric Halogen Compounds from a Satellite Using Gas Filter Spectroscopy in the Solar Occultation Mode," Appl. Opt., 16, 607-612, 1977.

2. Use of trademarks or names of manufacturers does not constitute an endorsement of such products by NASA.

3. Moore A. S., Mauldin L. E., Stump C. W., Fabert M. G., and Regan J. A., "Calibration of the Halogen Occultation Experiment Sun Sensor," Proc. of SPIE, 685-27, August 1986.

4. Mauldin L. E., Moore A. S., Stump C. W., and Mayo L. S., "Application of a Silicon Photodiode Array for Solar Edge Tracking in the Halogen Occultation Experiment," Proc. of SPIE, 572-23, August 1985.

5. Sullivan E. M. and Walthall H. G. "Fabrication of Glass Gas Cells for the HALOE and MAPS Experiments," NASA TM-86-86302, October 1984.

Calibration of the Halogen Occultation Experiment Sun sensor

A. S. Moore, L. E. Mauldin, C. W. Stump

NASA Langley Research Center, Hampton, Virginia 23665-5225

M. G. Fabert

System and Applied Science Corporation, Hampton, Virginia 23665

J. A. Reagan

Department of Electrical and Computer Engineering, University of Arizona,
Tucson, Arizona 85721

Abstract

The calibration of the Halogen Occultation Experiment (HALOE) Sun sensor is described. This system consists of two energy balancing silicon detectors which provide coarse azimuth and elevation control signals, and a silicon photodiode array which provides top and bottom solar edge data for fine elevation control. All three detectors were calibrated on a mountaintop in Tucson, Arizona, using the Langley plot technique. The conventional Langley plot technique was modified to allow calibration of the two coarse detectors which operate wideband. A brief description of the test setup is also given. The HALOE instrument is a gas correlation radiometer that is now being developed for the Upper Atmospheric Research Satellite (UARS).

Introduction

The HALOE instrument is presently being developed by NASA Langley Research Center for the Upper Atmosphere Research Satellite (UARS). HALOE is one of ten instruments on the UARS, which is scheduled for a late 1991 Shuttle launch. UARS is a free flying observatory intended for a circular orbit of 600 km altitude and 57 deg inclination.

The HALOE instrument is an infrared remote sensor that uses the solar occultation technique to scan the limb of the Earth during satellite sunrise and sunset events, as depicted in Figure 1. The atmospheric scan is achieved by tracking the Sun as it rises from the horizon to a tangent height of 65 km or tracking the Sun as it sets over the same distance. To employ this occultation technique, the instrument has a two-axes gimbal system that can accurately point the instantaneous field-of-view (IFOV) of the instrument to a given location on the solar disk. Tracking this location through the limb is also required. During the limb tracking, the instrument will measure the vertical distribution of seven upper atmospheric constituents: O_3, HC_1, HF, NO, CH_4, H_2O, and NO_2. In addition, an eighth channel (CO_2) is used to obtain the atmospheric pressure profile. The data collected will enhance the understanding of upper atmospheric chemistry through the determination of sources and sinks, validation of atmosphere dynamics models, and study of upper-atmosphere transport mechanism. The HALOE latitudinal coverage is from 75 deg S to 75 deg N, altitude coverage is from 10 to 65 km tangent height, vertical resolution is 2 km in all channels, and midstratosphere data inversion accuracy is 10 to 15 percent of full-scale at exoatmospheric. The acquisition of this data can best be accomplished with Sun sensors having a wide dynamic intensity range. Calibration of these sensors to exoatmospheric solar radiation levels is one of the most important tasks performed during instrument development. This paper describes a novel technique which was used to calibrate the HALOE Sun sensors.

HALOE Instrument Description

The HALOE instrument is shown in Figure 2 and an optical schematic is given in Figure 3. It contains four conventional radiometric channels and four gas correlation channels, all measuring between 2.43 and 10.25 micrometer wavelengths. The basic instrument characteristics are given in Table 1.

A rectangular field stop at the focal point of a 16 cm diameter Cassegrain telescope determines the HALOE 2 x 6 arcmin IFOV. A reflective dual track chopper wheel positioned at an image of the field stop provides modulation of solar energy at 150 Hz and modulation of an internal blackbody radiometric reference at 300 Hz. The reference source is used via an automatic gain control (AGC) loop to prevent instrument related changes from contaminating solar energy measurements during the data event for the four gas correlation channels.

The optical beam is separated by the reflective chopper into a radiometer beam and a gas correlation beam. Beam splitters and dichroic filters further divide each of these beams into four channels. The four radiometer channels are demodulated and telemetered using standard signal processing techniques. Each of the four gas correlation channels has two paths, one of which contains a gas cell and one which

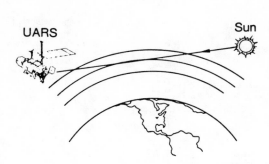

Figure 1. HALOE solar occultation experiment geometry

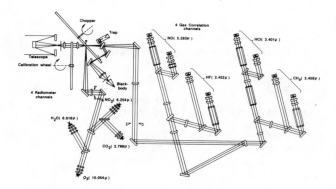

Figure 3. Optical schematic

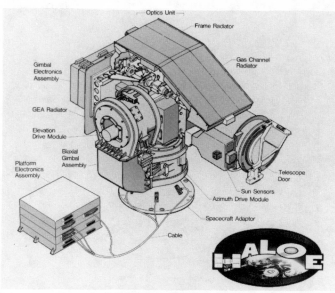

Figure 2. HALOE instrument

Table 1. Instrument characteristics

Telescope:	f/6.6, 96 cm fl Cassegrain
Wavelengths:	2.452, 2.799, 3.401, 3.459, 5.263, 6.254, 6.616, and 10.25 micrometers
Detectors:	InAs, HgCdTe, and bolometer
IFOV:	2 arcmin El by 6 arcmin Az
Spectral resolution:	20 to 120 cm^{-1}
Vertical resolution:	2 km at limb
Horizontal resolution:	6.3 km at limb
Instrument mass:	112 kg
Average power:	110 watts
Data rate:	4000 bps

does not. The AGC loop forces the detector signals from each path to balance while viewing the exoatmospheric Sun. The difference in output from each path is measured during the occultation event. When the target gas is present in the atmosphere, a spectral content is introduced into the incoming solar energy. This is correlated with the absorption line spectrum of the identical gas in the gas cell.

A stepper-motor driven calibration wheel, which is located after the field stop on the recollimated beam of the telescope, provides measurements of modulation scale factor and signal linearity using the exoatmospheric Sun as an energy source. The calibration wheel contains eight gas filled cells and three neutral density filters for these measurements.

The HALOE instrument operates virtually autonomously once powered and initialized after launch. Once commanded to the operate mode, the instrument performs a solar acquisition, balance, solar scan, calibration, track, and stow sequence as shown in Figure 4. The instrument automatically alternates between the sunrise and sunset sequence unless commanded otherwise. Two bi-level and sixteen 16-bit serial digital commands are used for instrument initialization and routine operation. The data system digitizes the science and engineering data (12-bit quantization) into a NRZ-L coded serial PCM telemetry stream of 4000 bits/sec. Additionally, 22 analog housekeeping measurements and 7 bi-level status checks are provided to the spacecraft interface for telemetry.

The HALOE instrument is pointed by coarse and fine Sun sensors and stepper-motor driven gimbals in a microprocessor based closed-loop feedback control system as shown in the block diagram in Figure 5. The Biaxial Gimbal Assembly (BGA) contains independently controlled azimuth and elevation gimbals with a 370 deg azimuth range and a 39 deg elevation range. Acquisition and tracking control signals for the gimbals are derived from Sun sensors which can acquire the Sun over a $\pm$ 3 degree field of view. The azimuth-axis uses an analog coarse Sun sensor (CSS) to acquire and track the azimuth radiometric centroid of the solar disk during a data taking event. The elevation-axis contains an identical analog coarse Sun sensor for solar acquisition and a digital fine Sun sensor (FSS) for solar scanning and edge tracking. The scan mode is used to obtain limb darkening data on the exoatmospheric Sun, while the track mode is used to obtain data during solar occultation. A more detailed instrument description can be found in Reference 1. The

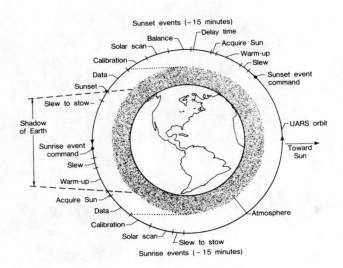

Figure 4. Orbital events summary

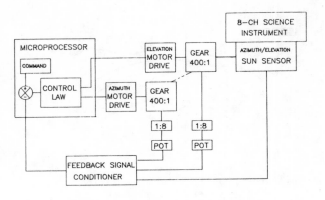

Figure 5. Pointer/tracker control system

Table 2. Fine Sun sensor design characteristics

Telescope:	f/12.5, 318 mm fl Galilean
Field-of-view:	+ 0.6 deg
Operating wavelength:	0.7 micrometer
Dynamic range:	100:1
Resolution:	16.2 arcsec
Detector:	256 element linear photodiode array

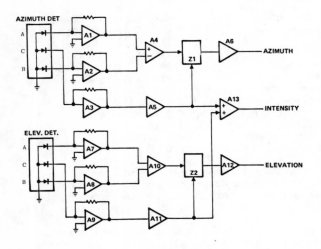

Figure 6. Coarse Sun sensor block diagram

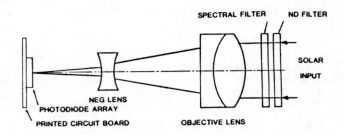

Figure 7. Fine Sun sensor optical characteristics

following paragraphs will briefly describe the CSS and FSS design, provide a detailed description of the exoatmospheric solar radiometric calibration techniques, test setup, and results.

CSS and FSS design

The CSS and FSS provide pointing knowledge to within 15 arcseconds in azimuth and 12 arcseconds in elevation, respectively. Two of the goals of the Sun sensor design are to have the CSS operate within 90 percent of its full-scale output and the FSS operate within 80 percent of its full-scale output. Achievement of these goals is the objective of the calibration described in this paper.

Each of the two CSS detectors consists of a three-element silicon photodiode arrangement located behind two masks. Three diodes of each detector are shown schematically in Figure 6. They are identified as A and B for the outer diodes and as C for the center. The solar rays illuminate one of the outer diodes (A) more than the other outer diode (B) unless the solar beam is aligned with the masks. Thus, the detector output of A minus B provides a signal proportional to angle and the output of C provides a signal proportional to solar irradiance. The geometry of the masks and diodes allows the detectors to provide angular and solar irradiance information.

The electronics for each CSS detector are similar enough that only one will be explained. The azimuth detector outputs are processed by transimpedance amplifiers A1, A2, and A3, as shown in Figure 6. Outputs of A1 and A2 are subtracted in A4 to form the difference signal which is proportional to a relative angle.

The angular output is derived from analog divider Z1, which forms the ratio of the outputs A4 and A5. Amplifier A6 contains a low-pass filter and serves as an output buffer. The processed signal of the center detector is summed with the other center detector to provide solar irradiance information at A13. This information is used for solar acquisition, gimbal pointing control and is also telemetered as pointing knowledge for data reduction.

The FSS consists of a photodiode array mounted at the focal point of the telescope depicted in Figure 7. Design characteristics are given in Table 2. The optical system consists of a diffraction limited f/12.5, 318 mm focal length Galilean telescope which is designed for monochromatic performance at 0.7 micrometer. An interference filter with a 30 Angstrom bandwidth sets the 0.7 micrometer operation wavelength. The telescope housing is made out of invar to maintain optical integrity over a wide instrument temperature range. This telescope, which is boresighted to the main HALOE telescope, images the solar disk on the linear photodiode array. This electronically self-scanned array contains 256 Si photodiode elements with a 25 micrometer center-to-center spacing, so that the solar image covers 118 elements. The HALOE IFOV is boresighted to photodiode element number 140 which places the IFOV 4 arcminutes below the top edge of the solar disk. A radiometric dynamic range of 100:1 provides altitude coverage down to a tangent height of 5 km. A neutral density filter located in front of the spectral filter is used to prevent the photodiode array elements from becoming saturated.

The electronics for the FSS is mostly digital. The analog video output from the photodiode array is converted to a binary video output as the array is scanned. Each element of the array that has an analog output greater than $1/3$ V_p is represented by a "1", the other elements are represented by a "0". V_p is the "average" output of the elements that are illuminated by the solar disk. This digital output is scanned for transitions from "0" to "1" and "1" to "0" which represent the solar top and bottom edges, respectively. The address of the element that represents the top solar edge provides a feedback signal for pointing the elevation gimbal during solar tracking and scanning. Correlation signals for the top and bottom solar edges are performed by the electronics to provide health status of the element in the array. V_p, solar edges, and correlation information are telemetered for data reduction. A more detailed description of the optical and electronic design of the FSS can be found in Reference 2.

The operating range for the CSS is adjusted electronically while the FSS is adjusted optically. A margin of 10 percent is reserved for calibration uncertainties for the CSS while 20 percent is reserved for calibration uncertainties and in-orbit filter degradation for the FSS.

Langley plot calibration technique

The directly transmitted solar spectral irradiance F_λ(wm^{-2}nm^{-1}) at wavelength λ received at the Earth's surface may be related to the exoatmospheric or zero-airmass solar spectral irradiance $F_{O\lambda}$ by the Bouguer-Lambert law.

$$F_\lambda = F_{O\lambda} T_\lambda \tag{1}$$

and

$$T_\lambda = e^{-m(\tau_\lambda + \tau_{\lambda}g)} \tag{2}$$

where

$\quad$ m $\quad$ is the atmospheric relative airmass for the solar position (zenith angle) at the time of observation of F_λ,

$\quad$ τ_λ $\quad$ is the atmospheric spectral optical depth at λ excluding any gaseous spectral absorption,

$\quad$ $\tau_{\lambda}g$ $\quad$ is the atmospheric gaseous spectral absorption optical depth,

$\quad$ T_λ $\quad$ is the atmospheric spectral transmittance at λ for airmass m and total optical depth $\tau_\lambda + \tau_{\lambda}g$, and

$\quad$ $F_{O\lambda}$ $\quad$ is the exoatmospheric solar spectral irradiance at λ for the Earth-Sun separation distance at the time of observation.

To obtain a measurable signal with the Sun sensor systems, F_λ must be collected over some non-zero wavelength interval and receiver area. The collected flux is also converted to a voltage by the system's photodetectors and electronic amplifiers. For a relatively small collection area A_R and narrow bandpass $\Delta\lambda$ centered on λ, the quasi-monochromatic system output voltage V_λ is given by

$$V_\lambda = A_R \int_{\Delta\lambda} R_\lambda F_\lambda \, d\lambda \tag{3}$$

where R_λ is the system responsivity including the spectral dependence of both the photodetector and the wavelength/bandpass selection system.

If $\Delta\lambda$ is fairly small, say less than about 10 nm, and λ is in a region excluding any significant gaseous spectral absorption so that $\tau_{\lambda}g \simeq 0$, then Eq. (3) is well approximated by

$$V_\lambda = V_{O\lambda}\, e^{-m\,\tau_\lambda} \qquad\qquad (4)$$

and

$$V_{O\lambda} = A_R \int_{\Delta\lambda} R_\lambda\, F_{O\lambda}\, d\lambda \qquad\qquad (5)$$

where $V_{O\lambda}$ is the exoatmospheric signal level. In this case, the surface signal V_λ is of the same form and, thus, obeys the Bouguer-Lambert law. Given that τ_λ is spatially/temporally constant over the range of airmasses that measurements are made, a plot of $\ln V_\lambda$ versus m, a Langley plot, will yield a set of data points distributed along a straight line with slope $-\tau_\lambda$ and intercept $\ln V_{O\lambda}$. The Langley method of determining τ_λ and $V_{O\lambda}$ by straight-line fitting Langley plot data has been used extensively to monitor atmospheric turbidity and estimate the exoatmospheric solar irradiance (e.g., References 3-5). This is the approach that was used to establish the exoatmospheric signal level for the HALOE FSS which has an interference filter defined bandpass of 3 nm centered at 700 nm. A modified approach described later was used to calibrate the CSS.

Test setup

The HALOE Sun sensor calibration tests were conducted on Mount Lemmon (elevation 2790 m) near Tucson, Arizona, although some backup measurements were also made in Tucson. A high altitude dry, clean site such as Mount Lemmon offers the advantage of minimizing atmospheric attenuation, thereby yielding a surface signal closer to the exoatmospheric level. The test setup is depicted in Figure 8. The HALOE Sun sensor was mounted beside a 10 channel standard radiometer. This radiometer, which was developed at the University of Arizona, consists of 10 narrowband (less than 10 nm) filters which are mounted in a filter wheel that can insert each filter sequentially in front of a silicon photodiode detector. The detector, which is temperature controlled to 40°C, had been previously calibrated by the University of Arizona (within 1 percent) in each of the 10 channels using the Langley plot technique. The center wavelength in 9 of the 10 channels was chosen to coincide with non-gaseous spectral absorption regions over the wave-length range from 370 to 1030 nm. The standard radiometer was used to evaluate atmospheric conditions for the test period and also for data analysis purposes in calibrating the HALOE coarse Sun sensors.

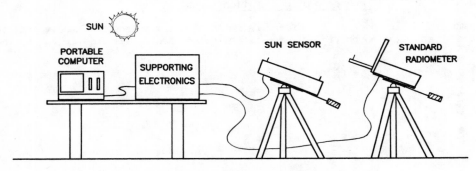

Figure 8. Sun sensor test setup

Pointing accuracy requirements were 1 degree for the standard radiometer and 0.25 degree for the Sun sensor. Alignment of the standard radiometer was accomplished manually by aligning a shadowed target. Alignment of the HALOE Sun sensor was accomplished by manually pointing the device until azimuth and elevation error signals (monitored with analog null meters) were within acceptable limits.

A supporting electronic system was used to provide all electronic interfaces to the instruments. A portable computer provided the following functions: (1) 16 channel data acquisition system; (2) time correlation to GMT; (3) Sun sensor alignment monitor; (4) prompts to begin data sequence; and (5) data processing. A typical data sequence was conducted as follows: The alignment techniques described previously were used to align the two instruments. At the sound of the computer prompt, the alignment accuracy from the computer monitor was checked and then the data sequence was initiated. Since the standard radiometer consisted of a single radiometric channel and 10 filters, prompts were issued by the computer every 6 seconds to the operator for rotating the filter wheel. Also, temperature information was provided on the computer monitor so the instrument temperatures could be maintained as the ambient temperature of the outdoor environment increased.

The menu driven computer data acquisition system was written in FORTH. The 16 channels were allotted as follows: 1 channel for standard detector output, 12 channels for the Sun sensor outputs, and 3 channels for instrument temperatures. Sun sensor outputs consisted of A, B, C, A-B, and (A-B)/C for both azimuth and elevation channels. The other two outputs, CSS magnitude and FSS magnitude, V_p ($V_{O\lambda}$, mentioned above, equals V_p when the FSS is pointing to the exoatmospheric Sun), were the channels used for calibration purposes.

At the conclusion of each 3-hour test, which consisted of approximately 35 separate data runs, the raw data were dumped and the computer was initialized for data processing. The data processing algorithms, which were written in FORTRAN, essentially provided the data analysis described earlier. The processed data were first used to evaluate atmospheric stability during the test. Then, data from the acceptable days were used for calibration purposes.

Results for FSS

The Langley plot for one set of Mount Lemmon observations (4/19/85) with the HALOE FSS is shown in Figure 9. It can be seen that the data points display almost a perfect straight-line behavior. However, such behavior does not in itself guarantee that the optical depth was constant and, thus, that the straight-line intercept determination is valid (e.g., References 6-8). Concurrent measurements were made with the standard radiometer to verify that the optical depth remained nearly constant over the observation period. Estimates of the HALOE FSS exoatmospheric signal level obtained from three days of observations (two days on Mount Lemmon and one in Tucson) are listed in Table 3. The value of $V_{O\lambda}$ obtained for each observation set (day) was obtained from a weighted least-squares straight line fitting procedure (Reference 9) applied to the Langley plot data for that day. While these $V_{O\lambda}$ values are relative in the sense that they cannot be absolutely related to a known irradiance level, they can be related to the full-scale signal range of the HALOE FSS for the system gain setting at the time of the observations. The design goal is to reset the system gain so that the $V_{O\lambda}$ is at 80 percent of the full-scale signal range. The disparity in the $V_{O\lambda}$ determinations given in Table 3 is sufficiently small (spread of less than ± 1 percent) to assure that a 20 percent margin in the signal range should be quite sufficient to avoid saturation of the HALOE FSS when it is above the atmosphere and pointed at the Sun.

Table 3. HALOE fine sensor system exoatmospheric signal levels
determined by Langley plot fits

Date and Location of Measurements	Estimated Exoatmospheric Signal Level[*] $V_{O\lambda}$ (relative counts)
4/19/85 – Mount Lemmon	8983
4/20/85 – Mount Lemmon	9021
4/23/85 – Tucson	9159
Average =	9054 ± 89

[*]Signals normalized to the mean Earth-Sun separation distance

Results for CSS

It was not possible to use the Langley method to determine the exoatmospheric signal levels for the HALOE elevation and azimuth CSS's because of their wide spectral sensing range. These CSS's have wideband spectral channels set by the spectral responsivity of the silicon photodiodes used for the system photosensors. The Bouguer-Lambert law, when integrated over a broad non-absorbing spectral region or even a narrow fairly weak gaseous spectral absorption region, does not yield a simple exponential relation like Eq. (4) and, hence, does not yield a linear Langley plot. For such situations, specially determined band transmittances must be used to extrapolate surface signals to equivalent exoatmospheric levels.

For wideband applications where $\Delta\lambda = \lambda_2 - \lambda_1$ extends over a wide spectral range (i.e., 100 nm or more), Eq. (3) may be expressed in the form

$$V = V_O \hat{T} \qquad (6)$$

and

$$V_O = A_R \int_{\lambda_1}^{\lambda_2} R_\lambda F_{O\lambda} \, d\lambda \qquad (7)$$

$$\hat{T} = \frac{A_R}{V_O} \int_{\lambda_1}^{\lambda_2} R_\lambda \ F_{O\lambda} \ T_\lambda \ d\lambda \tag{8}$$

where $\hat{T}$ is the wideband atmospheric transmittance. In spectral regions where $\tau_{\lambda g} = 0$, the spectral transmittance T_λ reduces to

$$T_\lambda = e^{-m\tau_\lambda} \tag{9}$$

but for even a small spectral interval $\Delta\lambda$ in a gaseous spectral absorption region T_λ takes the form

$$T_\lambda = e^{-m\tau_\lambda} \ \hat{T}_{\lambda g} \tag{10}$$

and

$$\hat{T}_{\lambda g} = \frac{A_R}{V_{O\lambda}} \int_{\Delta\lambda} R_\lambda \ F_{O\lambda} \ e^{-m\tau_{\lambda g}} \ d\lambda \tag{11}$$

where $\hat{T}_{\lambda g}$ is the spectral band-weighted transmittance determined by the particular spectral absorption behavior of the gas in question and the spectral features of $F_{O\lambda}$ and R_λ. Even if $F_{O\lambda}$ and R_λ are effectively constant over $\Delta\lambda$, the strong spectral variation of $\tau_{\lambda g}$ typically exhibited by gaseous absorption lines is sufficient to require the bandweighted spectral transmittance $\hat{T}_{\lambda g}$.

Determination of $\hat{T}$ requires knowledge of the exoatmospheric spectral irradiance $F_{O\lambda}$, the relative spectral variation of the system responsivity R_λ, and T_λ. The NASA standard developed by Thekaekara (Reference 10) was used for $F_{O\lambda}$, and R_λ was taken as the spectral responsivity of a silicon photodiode. Two silicon photodiode responsivity curves were considered. One was a nominal silicon responsivity curve and the other was a blue enhanced. The products of $F_{O\lambda}$ times R_λ for each of these responsivities (denoted as NOM or BLU ENH) are plotted versus wavelength in Figure 10. It can be seen that the BLU ENH responsivity weighs shorter wavelengths significantly more than nominal responsivity. The areas under these two curves were summed numerically, in 10 nm steps, to yield relative exoatmospheric signal levels, V_O, as required in Eq. (7).

The next step in computing $\hat{T}$ is to determine T_λ, as defined in Eq. (10), so that the product $R_\lambda F_{O\lambda} T_\lambda$ can be evaluated for the numerator integrand of Eq. (8). This requires determining τ_λ and, for regions where gaseous absorption occurs, $\hat{T}_{\lambda g}$. Since τ_λ is due to components that vary smoothly with wavelength, namely, molecular (Rayleigh) scattering component, the aerosol extinction component, and the continuum gaseous absorption component, τ_λ may be estimated fairly accurately by extrapolating between values of τ_λ determined at only a few wavelengths.

To obtain these few required values of τ_λ, concurrent measurements were made with the standard radiometer with nine channels positioned in non-gaseous spectral absorption regions over the wavelength range 370 to 1030 nm. The τ_λ values were determined by straight-line fitting the Langley plot data for each radiometer channel in the same manner as described earlier for the HALOE fine Sun sensor. The results obtained for one of the Mount Lemmon observation days (4/19/85), including the smooth curve extrapolated through the τ_λ data points, are plotted in Figure 11. The uncertainty in the τ_λ values obtained from the Langley analysis is estimated to be less than $\pm$ 0.005.

The only gaseous spectral absorption of any real consequence for the silicon photodiode bandpass occurs due to water vapor absorption bands around 720, 840 and 940 nm, and oxygen absorption bands around 760 nm. The gaseous spectral band transmittances $\hat{T}_{\lambda g}$ for each of these bands were computed using the LOWTRAN VI computer code (Reference 11). Pressure, temperature and dew point temperature profile information required for the LOWTRAN computations were obtained from weather station radiosonde measurements made over Tucson on the same days that the HALOE Sun sensor measurements were made.

For $\hat{T}_{\lambda g}$ computed for a given airmass m (i.e., for a given observation time) and for τ_λ determined as discussed in reference to Figure 11, T_λ may be accurately computed according to Eq. (10) for small $\Delta\lambda$ where τ_λ is evaluated at the center of $\Delta\lambda$. Using T_λ determined in this manner, the products of $R_\lambda F_{O\lambda} T_\lambda$ were computed at selected times for each observation day. Results are shown in Figures 12, 13, and 14 for two days of observations on Mount Lemmon and one day in Tucson. As with Figure 10, the computations were made for both nominal and blue enhanced responsivities. The areas under the curves in these figures

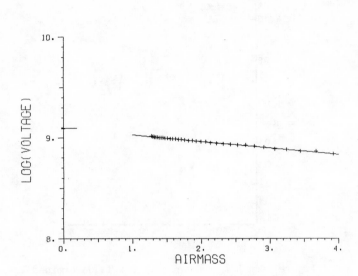

Figure 9. Langley plot of HALOE FSS observations made at Mount Lemmon, AZ on April 19, 1985, from 7:02 to 10:01 MST

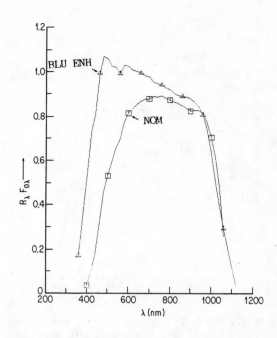

Figure 10. $R_\lambda F_{O\lambda}$ versus λ computer for NOM and BLU ENH responsivities

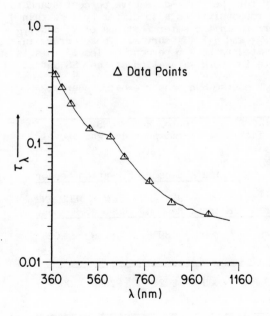

Figure 11. τ_λ versus λ determined from spectral solar radiometer observations at Mount Lemmon, AZ on April 19, 1985, from 7:02 to 10:01 MST

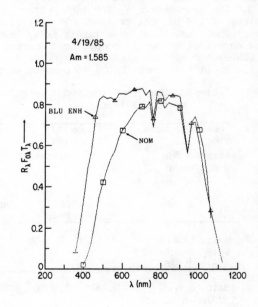

Figure 12. $R_\lambda F_{O\lambda} T_\lambda$ versus λ for T_λ determined for April 19, 1985, 9:00 MST Mount Mount Lemmon, AZ solar observations

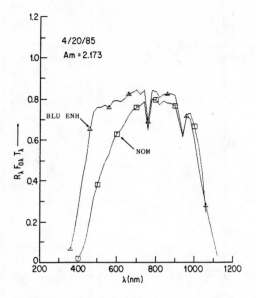

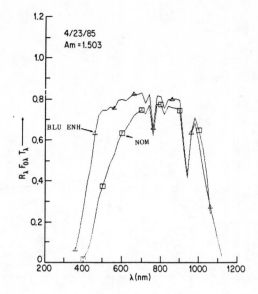

Figure 13. $R_\lambda F_{O\lambda} T_\lambda$ versus λ for T_λ determined for April 20, 1985, 8:02 MST Mount Lemmon, AZ solar observations

Figure 14. $R_\lambda F_O \, T_\lambda$ versus λ for T_λ determined for April 23, 1985, 16:53 MST Tucson, AZ solar observations

were summed numerically, in 10 nm steps as was done for Figure 10, to yield relative surface signals, V. The wideband transmittances $\hat{T}$ were then determined by dividing these areas by the areas determined for the $R_\lambda F_{O\lambda}$ curves in Figure 10. The resulting values of $\hat{T}$, the actual HALOE CSS surface signal observations at the same times, and the computed HALOE CSS exoatmospheric signal levels (computed by Eq. 6) are given in Table 4. The averages of V_O for each sensor and responsivity for the three days have percent standard deviations of 1.4 percent or less. The averages for the two responsivities also differ by less than 4 percent. Some additional uncertainty in V_O arises due to errors in the determinations of τ_λ and $\hat{T}_{\lambda g}$ as well as numerical errors in computing the areas under the $R_\lambda F_{O\lambda}$ and $R_\lambda F_{O\lambda} T_\lambda$ curves. These errors in the HALOE CSS exoatmospheric signal level determination are estimated to be $\pm$ 5 percent or less. This is sufficiently small to assure that a 10 percent margin is adequate to avoid saturation of the CSS detectors.

Table 4.   Computed wideband transmittances and HALOE coarse Sun sensor exoatmospheric signal levels

Location, Date, Time, and Airmass for Measurement	Computed Wideband Transmittance $\hat{T}$		Observed HALOE CSS Sun Sensor Surface Signal Level* V		Computed HALOE CSS Exoatmospheric Signal Level* V_O			
					Azimuth Sensor		Elevation Sensor	
	NOM Respon.	BLU ENH Respon.	Azimuth Sensor	Elevation Sensor	NOM Respon.	BLU ENH Respon.	NOM Respon.	BLU ENH Respon.
Mount Lemmon 4/19/85–9:00 MST m = 1.585	0.8728	0.8469	7585	7645	8690	8956	8759	9026
Mount Lemmon 4/20/85–8:02 MST m = 2.173	0.8419	0.8060	7274	7333	8639	9024	8709	9097
Tucson 4/23/85–15:36 MST m = 1.503	0.8076	0.7759	7131	7192	8831	9192	8905	9269
					Avg. of 8720 $\pm$99	Avg. of 9057 $\pm$121	Avg. of 8791 $\pm$102	Avg. of 9131 $\pm$125

*Normalized to mean Earth–Sun separation distance

Conclusions

The Langley plot technique was used to calibrate the HALOE narrowband fine Sun sensor to its exoatmospheric solar output to within 5 percent. A novel technique was developed, using a modified Langley plot approach, to calibrate the HALOE wideband coarse Sun sensors to within 5 percent. Results of the data analysis were used to select a neutral density filter to set the fine Sun sensor exoatmospheric output to 80 percent of its full scale output and to select gain resistors to set the coarse Sun sensor exoatmospheric output to 90 percent of its full scale output. Also developed for this application was a novel computer software package using FORTH programming language to automate, monitor, and control the test for the test conductor. The techniques developed should have future application for calibrating similar Sun sensors.

Acknowledgements

The original design of this device was done by TRW, Inc. under contract NAS1-15880. Design, development, fabrication, assembly, alignment, and calibration was accomplished by the HALOE team at NASA Langley, which is managed by J. Raper for the Principle Investigator, J. Russell. The authors would like to thank those at NASA Langley and University of 'rizona, especially I. C. Scott-Fleming for his computer graphics support, who spent many hours in preparing for and supporting the calibration of the HALOE Sun sensor. Also, a special thanks goes to A. S. Rockey for her professional secretarial skills in preparing this paper.

References

1. Baker, R.L., L.E. Mauldin, and J.M. Russell, "Design and Performance of the Halogen Occultation Experiment (HALOE) Remote Sensor," Proc. of SPIE, 685-26, Aug. 1986.

2. Mauldin, L.E., A.S. Moore, C.W. Stump, and L.S. Mayo, "Application of a Silicon Photodiode Array for Solar Edge Tracking in the Halogen Occultation Experiment," Proc. of SPIE, 572-23, Aug. 1985.

3. Ångstrom, A., "On the Atmospheric Transmission of Sun Radiation and On Dust in the Air," Geogr. Ann., Vol. 11, pp. 156-166, 1929.

4. Shaw, G.E., J.A. Reagan and B.M. Herman, "Investigations of Atmospheric Extinction Using Direct Solar Radiation Measurements Made With a Multiple Wavelength Radiometer," J. Appl. Meteor., Vol. 12, pp. 374-380, 1973.

5. Reagan, J.A., L.W. Thomason, B.M. Herman, and J.M. Palmer, "Assessment of Atmospheric Limitations on the Determination of the Solar Spectral Constant from Ground-Based Spectroradiometer Measurements," IEEE Trans. Geosci. Remote Sensing, Vol. GE-24, pp. 258-266, 1986.

6. Young, A.T., "Observational Technique and Data Reduction," Optical and Infrared Methods of Experimental Physics, Vol. 12: Astrophysics, Part A., N. Carleton, Ed. New York: Academic Press, pp. 123-192, 1974.

7. Shaw, G.E., "Error Analysis of Multi-Wavelength Sun Photometry," Pageoph, Vol. 114, pp. 1-14, 1976.

8. Reagan, J.A., I.C. Scott-Fleming, B.M. Herman and R.M. Schotland, "Recovery of Spectral Optical Depth and Zero-Airmass Solar Spectral Irradiance Under Conditions of Temporally Varying Optical Depth," in Proc. IGARSS '84 Sym. Strasbourg, August 27-30, 1984, pp. 455-459, ESA Scientific and Technical Publications Branch, Ref. ESA SP-215.

9. Herman, B.M., M.A. Box, J.A. Reagan and C.M. Evans, "An Alternate Approach to the Analysis of Solar Photometer Data," Appl. Opt., Vol. 20, pp. 2925-2928, 1981.

10. Thekaekara, M.P., "Extraterrestrial Solar Spectrum, 3000-6100 A at 1 A Intervals," Appl. Opt., Vol. 13, pp. 518-522.

11. Kneizys, F.X., E.P. Shettle, W.D. Gallery, J.H. Chetwynd, Jr., L.W. Abreu, J.E.A. Selby, S.A. Clough, and R.W. Fenn, Atmospheric Transmittance/Radiance: Computer Code LOWTRAN 6, AFGL-TR-83-0187, Air Force Geophysics Laboratory, Hanscom AFB, MA, 1983.

Novel long path transmissometry

S. T. Hanley, B. L. Bean, and R. Soulon

OptiMetrics, Inc.
106 E Idaho, Suite G, Las Cruces, NM 88005

J. Randhawa and R.A. Dise

U.S. Army Atmospheric Sciences Laboratory
White Sands Missile Range, NM 88002

Abstract

Novel transmissometry techniques developed for use in support of long path extinction measurements are described along with samples of data from natural and manmade obscurants. Advantages derived from these new techniques include precise control over field of view (FOV), ultra-low drift rates in system baseline response, identical alignment of all multispectral wavelength bands, and negligible humidity and temperature effects on system performance. A multispectral differential field of view technique developed for use with the U.S. Army Atmospheric Sciences Laboratory's SMART system is described along with measurement results. The differential field of view results have implications on sharply peaked forward scatter phase functions in the presence of obscurants from visible to 14-μm wavelengths.

Introduction

Long path atmospheric propagation measurement techniques have been under development at the U.S. Army Atmospheric Sciences Laboratory (ASL) and OptiMetrics, Inc. Among the more difficult aspects of atmospheric transmission measurement are system stability, calibration, and knowledge of and control over system operating parameters such as field of view, optical bandpass, and linearity. Solutions to some long path propagation measurement problems are presented in this work, along with samples of measurement results.

Optical System

Efforts to measure propagation at multiple wavelengths simultaneously through a common atmospheric path have led to the development of the current optical configuration used in the ASL Simultaneous Multispectral Absolute Radiometer and Transmissometer (SMART) system. The SMART system includes a source van and a receiver van, each with optical systems described below.

Source Optical System

Figure 1 shows optical components comprising the source system. The number of optical surfaces used has been kept to a minimum to reduce unwanted wavelength sensitivities within desired optical bandpasses, to minimize polarization effects, to reduce optical throughput losses and degradation from accumulation of dust out of doors, and to reduce labor expended in recoating optical surfaces.

Blackbody Source. The regulated blackbody source is used for broadband transmissometry measurements at all optical bands. Cavity temperature in the source is maintained at 1000 $\pm$5 °C, which is sufficient for good signal to noise from visible through 12-μm wavelengths. Selected apertures are positioned in front of the blackbody cavity to control the projected field of view from the source optical system.

Source Optical Chopper. The optical chopper is polished on both surfaces of the blade so that chopper temperature drift does not affect system response in the thermal bands, and reflection of blackbody radiation from the rear surface can be used to monitor source radiation level stability. When the chopper is in the closed position, it reflects the receiver optical field of view (from the polished rear surface) into a stop lined with a honeycomb light trap and fixed-temperature surface. The polished optical chopper is phase-locked to a crystal oscillator for stability and precisely known chopping frequency.

Laser Sources. Laser sources include CO_2, Nd:YAG, and HeNe. These lasers are combined with germaniun and zinc selenide flats into a collimated beam, and are inserted into the optical system, as shown in figure 1. The lasers may be used individually, or in any combination.

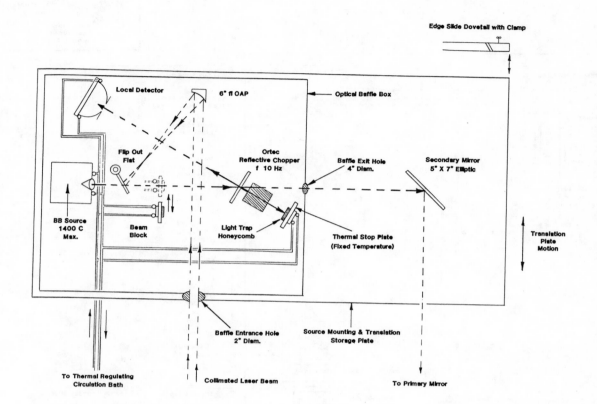

Figure 1. Source optical system configuration showing use of broadband blackbody and laser radiation sources.

Beam Block. A mechanical block, activated by remote control from the receiver van, is located between the source and polished chopper in the source van. This beam-blocking mechanism is surfaced with a honeycomb light trap and regulated in temperature to allow direct measurement of total extinction at any user-selectable time.

Primary Telescope. The primary mirror is a .9-m, f/5.5 parabola providing spatial averaging and good light-gathering power over longer path lengths (> 2 km). The entire optical train from source through Newtonian secondary is mounted on a translation plate with rails aligned to the optical axis of the primary mirror. In this way, internal alignment can be accomplished once with no need for realignment after focal adjustments. The focal adjustment is a simple movement of this translation plate with respect to the primary mirror.

Receiver Optical System

The receiver optical system is shown in figure 2. The design is similar to that of the source optical system, with a minimal number of optical surfaces, and a translation plate aligned to primary mirror optical axis. A polished optical chopper is used to alternately pass or direct the received energy into two detector assemblies. In the closed position, the chopper for each detector directs the field of view for that detector into a fixed-temperature plate lined with a honeycomb light trap. Each detector assembly utilizes a single field of view stop that is user-adjustable and fixes a common field of view for all optical channels within that assembly.

Optical Field Of View. Figure 3 illustrates the simplicity of the field of view control for each detector assembly. The field of view is easily adjusted to the desired value independently for each detector assembly according to the following equation

$$FOV \text{ (radians)} = [d(O-f)]/[(O)(f)], \tag{1}$$

where d is the field stop diameter, f is the focal length of the primary mirror, and O is the object range, and d, f, and O are all in the same units.

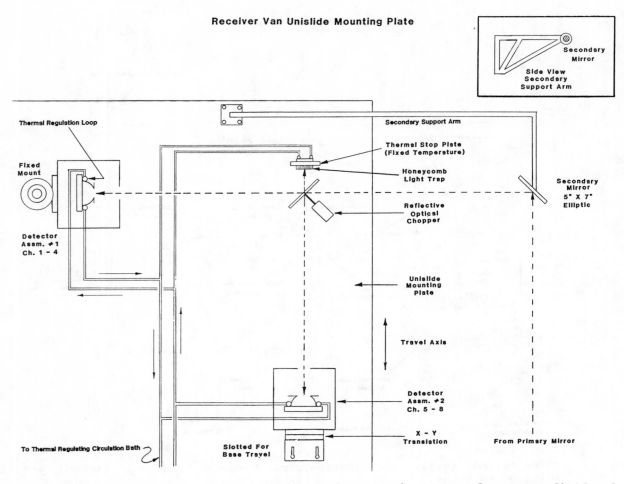

Figure 2. Receiver optical system configuration showing use of common field of view
detector assemblies and on-axis light trapping.

Detector Assemblies. The detector assembly that houses four discrete detectors and associated optical filters is shown in figure 4. Incident radiation passes through the field of view iris and impinges on the removable scatterplate. Four detectors are equally spaced azimuthally at 45 degrees from the incident axis. Scattered radiation subsequently passes through one of the four optical bandpass filters and is viewed by the corresponding detector. The aluminum housing has a highly polished, interior, hemispherical cavity, and is temperature-regulated for stability of detector resposivities. A Faraday shield encloses the detector assembly to reduce electromagnetic interference from such sources as lightning, walkie talkies, and radar. A dry nitrogen purge is used to flush out the hemisphere cavity volume and prevent dust or condensate from collecting on optical surfaces under windy or humid conditions.

Optical Bandpasses. Two complete detector assemblies are used for redundant measurement of transmission at the same optical bandpasses. Figure 5 shows the four optical bandpasses currently in use. Custom filter requirements can easily be exchanged for one or more of the optical bands currently in use.

Measurement Techniques

Measurement techniques have been developed that optimize detection of relatively small source modulations in the presence of backgrounds containing strong and widely varying on-axis flux levels, and that provide first-time measurement of concentric, differential field of view transmissions.

Phase-Locked Detection

All eight transmission channels are phase-locked to the source optical chopper. This provides maximum signal to noise in an otherwise noisy background radiation environment. The phase-locking across long path lengths is achieved with the aid of a 7 GHz fm microwave data link providing virtually distortion-free source chopper waveforms at the receiver

Optical System FOV

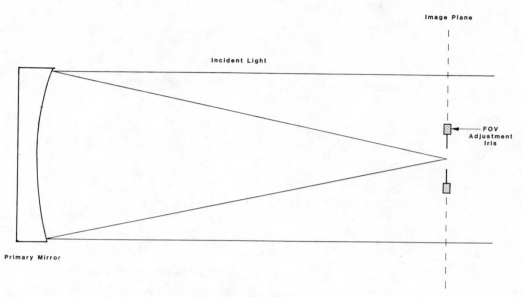

Figure 3. Optical system field of view showing single adjustable stop for all optical bands.

Optical Detector Assembly

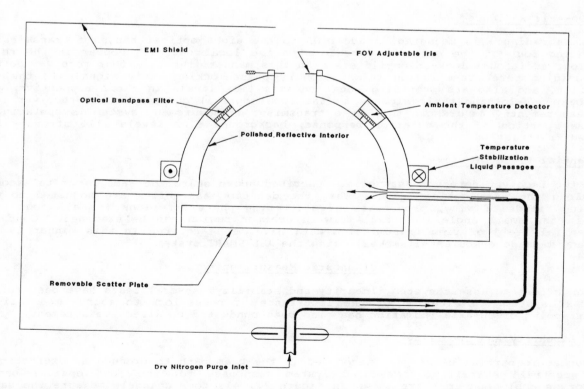

Figure 4. Receiver detector assemblies showing multiple optical bands and an adjustable common field of view.

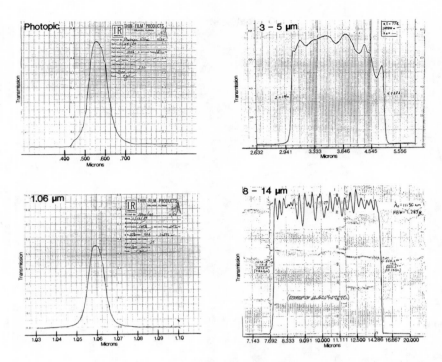

Figure 5. Optical bandpasses currently in use with double redundancy for each band.

position, as shown in figure 6. As will be discussed later in this paper, the phase-lock detection combined with trapping of on-axis background radiation forms the basis for novel precision in long-path outdoor transmission measurement.

Total On-Axis Radiance

On-axis radiance is measured in addition to the eight optical bands of transmission by referencing four of the existing detectors to the local optical chopper in the receiver using four additional lock-in amplifiers. In this manner, the same four receiver detectors are providing pure transmission information by phase-locking their signals to the source modulation, and also are providing total on-axis flux levels in their respective optical bands by phase-locking their signals to the local optical chopper in the receiver system. Past measurements[1] have shown that if a transmission measurement system is operating in a nonlinear portion of the detector response, background flux levels will strongly affect measured transmission.

Differential Field of View

The two receiver detector assemblies described above share the same large telescope and view along the same optical path. When the detector assemblies are adjusted to widely differing fields of view, properties of scattering phase functions for all four optical bands in the small angle ($\leqslant$5 mrad) forward scatter region can be examined. Concurrent, differential field of view transmission data have been acquired in this manner for snow, dust, and manmade obscurants (smokes) using the ASL SMART system.

Diagnostic Measurements

In an effort to assess system linearity and stability, several tests have been run using the SMART system. Evaluation of the importance of phase-lock to source modulation and light trapping of on-axis radiation has also been conducted by direct measurement.

System Linearity and Stability

A geometric partial block was introduced in the beam path to produce an extinction with total spectral neutrality. Measured system responses for the four optical bands in detector assembly number 1 are shown in figure 7. All four channels measure the same 44-percent transmission level, followed by total blockage, and a return to a 44-percent level, to better than 1 percent precision over a 500-m path during this diagnostic test. System stability is demonstrated by a lack of drift or wandering from channel to channel over this relatively short, clear air path.

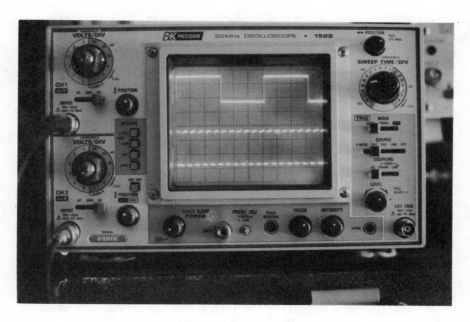

Figure 6. Reference waveform from source optical chopper (top trace) after 1.2-km transmission over 7 GHz FM data link, and reference waveform from local optical chopper (bottom trace) in receiver.

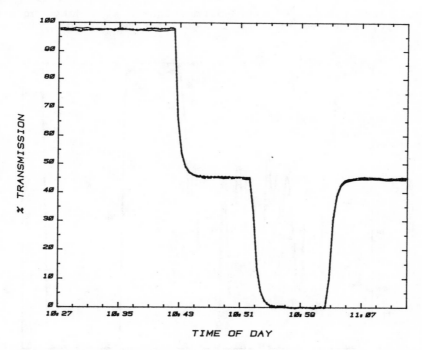

Figure 7. Diagnostic test showing transmission response of all four optical bands to 56 percent geometric (spectrally neutral) block. The sequence is full transmission (500 m), 44 percent transmission, total beam block, and 44 percent transmission, simultaneous to all channels.

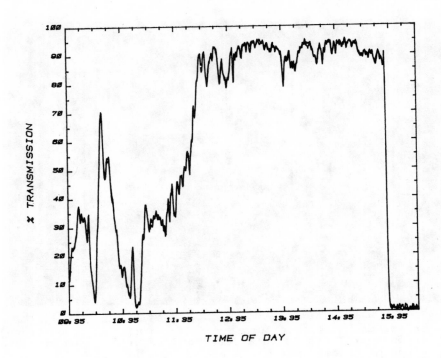

Figure 8.   Diagnostic test overlay of two 3-5 µm channels from detector assemblies 1 and 2, respectively, adjusted to same fields of view during dynamically varying snow attenuation.

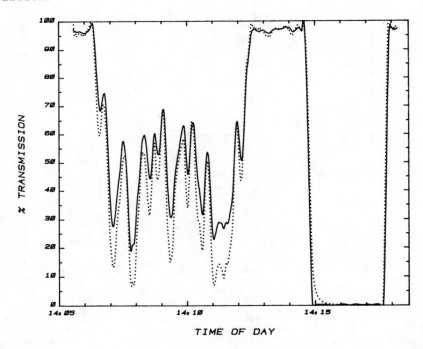

Figure 9. Measured dust obscuration in the visible optical band with phase-locked detection (solid curve), and with narrow-band (Q = 100) amplification tuned to source modulation without phase-locked detection (dotted curve).

In a second diagnostic test, system response to a snow storm with dynamically varying extinction over a 500-m path was examined. The two detector assemblies were adjusted to equal values of 2 mrad field of view, and the same 3-5 μm optical bands from each detector assembly were overlayed (figure 8). This diagnostic demonstrates very little system bias and good stability (typically better than 1 percent) over a wide range of extinction and for dynamically varying conditions.

Phase-Lock Versus Narrow Band

A diagnostic test was conducted over a 1-km path in which detector assembly number 1 was operated phase-locked to the source modulation, and detector assembly number 2 was operated with a narrow band amplifier tuned (with a Q of 100) to the source chopper frequency. Both contained the same four optical bands and were measuring the same optical path. The obscurant used was road dust over a 1-km total path length, with high solar loading. The results clearly show that phase-locking the received detection system is imperative in obtaining accurate extinction results (figure 9). The errors associated with narrow-band amplification techniques are rooted in the background flux levels during obscuration. Figure 10 shows the abrupt change in on-axis radiation at 1.06 μm during obscuration with the flux levels 180 degrees out of phase and mirroring the actual extinction.

Visible Versus 1.06-μm Radiation

Recent measurements with the novel measurement techniques described above have shown surprising results in the vicinity of 1.06-μm wavelength. Currently accepted mass extinction coefficients for fog oil, a screening aerosol in wide use within the DoD, predict that 1.06-μm radiation transmits through fog oil screens more readily than visible radiation. Contrary to these accepted relationships, when phase-lock and trapping of on-axis flux techniques are employed, 1.06-μm radiation is measured to be more strongly attenuated than visible for fog oil, as shown in figure 10(a). The same unexpected relationship of 1.06 μm to visible extinction also holds for certain other classes of obscurants.

Conclusion

The ASL SMART system offers some novel approaches to long path transmissometry. By combining the large optical telescopes with phase-lock detection across long paths, on-axis trapping of background flux, thermal regulation of surfaces within the field of view, common field of view control for all optical bands, and dual detection at all optical bands, the SMART system is able to provide a broad and accurate data base for atmospheric propagation research.

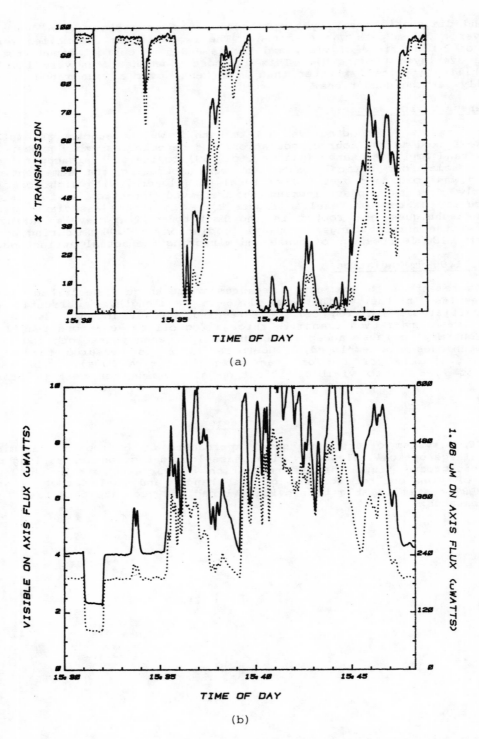

Figure 10. Measured transmission (a) and concurrent on-axis radiation (b) for fog oil obscurant, with visible radiation shown as solid curve and 1.06 μm radiation shown as dotted curve.

References

1. R. Davis, "The Effect of Background Light on Transmissometry: The XM-819 Developmental Trials," Smoke Symposium X, Harry Diamond Laboratories, Adelphi, MD, 1986.

Application of an infrared gas-filter correlation spectrometer for
measurement of methanol concentrations in automobile exhausts

Soyoung Cha
Department of Mechanical Engineering, The University of Illinois at Chicago
P.O. Box 4348, Chicago, Illinois 60680

Abstract

Spectroscopic methods provide an attractive alternative to wet chemical methods due to
their fast responses. An instrument using the principle of gas-filter correlation was built
in a laboratory and tested to measure methanol concentrations in exhausts from a methanol-
fueled vehicle. The instrument utilized the infrared spectrum between 8 μm and 11 μm. The
sensitivity and discrimination against other interfering gases were adequate enough to obtain
a detection limit of 0.5 ppm. The precision of the instrument varied from 5.5% to 1.2% for
methanol concentrations ranging 7 ppm to 113 ppm. The study demonstrated a good agreement
with the gas chromatograph analyses. The instrument has a strong potential for real-time
monitoring of automobile emissions. A small drift in the zero setting was primarily due to
the unstable temperature of optical components, especially the liquid-nitrogen cooled detector. An improvement in stability can be achieved by thermally insulating the analyzer to
maintain a constant temperature.

Introduction

Exhausts from methanol vehicles contain various pollutants including carbon monoxide (CO),
nitrogen oxides (NO_x), hydrocarbons (HC's), and oxygenated HC's. While CO and NO_x emissions
can be measured effectively with well-established optical methods such as nondispersive-
infrared and chemiluminescent analyzers, new efficient techniques need to be developed for
fast measurement of oxygenated HC's. The two oxygenated HC's of prime concern, emitted from
methanol cars, are methanol and formaldehyde. Although wet chemical procedures for measuring
these species have been available[1], the alternate optical method using a gas filter correlation spectrometer (GFCS) appears to be promising as compared with the tedious and time-
consuming wet chemical methods. By employing the GFCS, the oxygenated HC concentrations can
be determined rapidly and precisely[2] in a similar manner as the fore-mentioned optical
techniques for measuring CO, NO_x and HC's. References 3-11 describe some details of the
operation principle of the GFCS and the previous work results.

This paper discusses the methanol analyzer employing the gas-filter correlation principle
and its experimental results. The experiments were conducted by measuring the low concentration methanol vapor existing in exhausts from automobiles powered by a methanol-gasoline fuel
blend.

The GFCS studied utilizes a broad band spectrum of infrared radiation without spatially
separating various frequencies. Fundamentally, the system responds to the degree of
correlation between the two spectral absorptions of the gas being measured (object gas) and
the filter gas in the system. For these reasons, the instrument has several advantages
including high energy throughput good signal-to-noise ratio and low sensitivity to interfering gases as compared with dispersive-type infrared analyzers.

Description of the instrument and its operation

The schematic diagram of the methanol analyzer used for the experimental study is
illustrated in Figure 1. The components of the analyzer can be divided into an optical part
and an electronic part as shown in the diagram. The optical part includes a source, optical
filter, sample cell, beam splitter, chopper, gas-filter cell (GFC), detector and mirrors.
The electronic part consists of a detector preamplifier, two lock-in amplifiers, a ratiometer
a chopper control, a strip-chart recorder.

A Nernst glower of size 1.33 mm x 13 mm, operating at a temperature in excess of 1500k,
was used for the infrared source. An image of the source was formed near the inlet window
of the sample cell by a spherical concave mirror with a focal length of 33.5 mm. The source
radiation of wavelength below 8 μm was eliminated by using a long pass filter between the
sample cell and the concave mirror. The sample cell employed the multiple-pass optical
principle that was first described by White[12]. Its base length and volume were approximately
1.0 m and 10 ℓ, respectively. This long multiple-pass sample cell contained three spherical
concave mirrors all with the same focal length of about the sample-cell length. The source
image formed near the inlet window of the sample cell was reflected by a rear concave mirror
and reimaged on the front concave mirror. The front mirror reflected the radiation to the

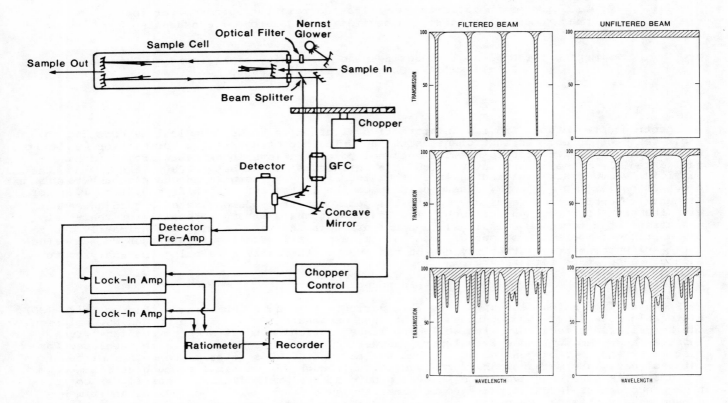

Figure 1. Schematic of GFCS methanol analyzer.

Figure 2. Absorption spectra in two beams

remaining rear concave mirror. This mirror also reflected the radiation to form the second image on the front mirror. This imaging process was repeated 20 times to produce an optical pathlength of about 36 m by adjusting the reflection angles of the rear concave mirrors. The long optical pathlength allowed for a substantial increase in the spectral absorption of the sample gas at low concentrations and thereby increased the analyzer sensitivity. While operating, the sample gas was continuously pumped through the cell at a flow rate of 15 ℓ/min. After exiting the sample cell, the beam was divided by a beam splitter. The two divided beams were then modulated by a two-channel lock-in chopper operating at frequencies of 39 Hz and 247 Hz. The beam modulated at 39 HZ was filtered through a small GFC 10 cm in length, that contained methanol vapor. For normal operation, the methanol pressure in the GFC was maintained at 122 torr. Both beams, filtered and unfiltered, were then reflected off two spherical concave mirrors with a focal length of 66 mm onto a HgCdTe detector that was cooled by liquid nitrogen. Barium fluoride windows were used for both the sample cell and GFC. The detector window was made of Irtran II (ZnS). The combined effects of the source, optical filter and windows allowed the use of the spectral band between 8 μm and 11 μm.

The detected intensities of the filtered and unfiltered beams, being modulated at two distinct frequencies, were amplified separately by lock-in amplifiers. Output signals from the amplifiers were directed to the ratiometer. The ratiometer provided a voltage output that was proportional to the ratio of outputs from the amplifiers-the signal from the filtered beam divided by that from the unfiltered beam. In the current study, the output of the ratio meter was kept constant at 1.0 volt when no sample was present in the sample cell. First, the two beams were adjusted to almost-equal intensities. The electronic amplification factors were then controlled to obtain the desired 1.0 volt output of the ratiometer.

The operation principle of the system can be easily understood by considering the spectra of the filtered and unfiltered beams as illustrated in Figure 2. The absorption spectra of the GFCS without the object gas in the sample cell are shown at the top of the figure. In this example, a continous spectrum of the unfiltered beam is assumed for simplicity. The filtered beam, that has passed through the GFC additionally, introduces the absorption lines into the continuous spectrum as shown in the diagram. Near the centers of the strong absorption lines, the transmittance of the GFC becomes very close to zero. Consequently, most of the energy of the filtered beam, that would be absorbed by the same gas (object gas) present in the sample cell, is removed by the GFC before it reaches the detector. In the diagram, the unfiltered beam has been adjusted to yield the total intensity equal to that of the GFC filtered beam. This adjustment can be easily accomplished optically by using a neutral density filter or trimmer, and electronically by controlling the lock-in amplifier or ratiometer. When no object gas is present in the sample cell, the analyzer will indicate a constant intensity ratio (1.0 volt) of the beams. The center of Figure 2 shows the absorption spectra of the two beams when the absorption lines of the gas in the sample cell match with those of the gas in the GFC. The addition of the object gas to the sample cell produces an appreciable decrease in intensity of the unfiltered beam, while it has a negligible effect on the intensity of the filtered beam. Now the detection system measures the increased intensity ratio indicating ratiometer output greater than 1.0 volt. The lower diagrams demonstrate the intensity changes when interfering gases are also present in the sample cell. If these gases have different spectral absorption lines than those of the object gas, then both beams are reduced almost the same proportion resulting in the negligible change in the instrument output.

In summary, when the sample cell contains the same gas species as that in the GFC, the strong spectral absorption correlation between the GFC gas and the object gas causes more attenuation in the unfiltered beam. The result of this measurable change in the beam intensity ratio can be related to the object gas concentration in the sample cell. However, if the sample cell contains different gas species, the negligible spectral absorption correlation results in almost the same intensity ratio of both beams. Interfering gases will increase the output when their absorption lines overlap with the lines of the object gas and will decrease the output when the lines do not overlap. If the interfering gases has many lines in the spectral region being used, the positive and negative interferences might nearly cancel each other. It is clear that continuous absorption, scattering particles or soiling of sample cell windows will not affect the balance between the two beams, since they reduce both beams at almost the same proportion. For the reasons stated above, the GFCS has a low sensitivity to interfering gases and particulates.

Obviously, proper choice of source, detector and filters will be very important to maximize the analyzer performance. The spectral band of the system should center around the strong absorption lines of the object gas. If the object gas has a broad band of fine absorption lines in a region relatively free from interferences, the optimum system design would utilize a spectral band slightly wider than the absorption band of the object gas. The spectral band 8 ~ 11 μm employed for the current study encompasses strong spectral absorption lines for methanol[13]. As discussed later in the experimental results, this range resulted in good sensitivity to low methanol concentrations and good discrimination against interfering gases.

In this study, the instrument response to a methanol gas unknown concentration was taken as the difference between the ratiometer reading of the sample minus that of the zero concentration (1.0 volt). The response of the instrument was monitored by using a strip-chart recorder.

Instrument characteristics and automobile exhaust measurements

Signal responses of the methanol analyzer were measured against a group of standard gases with different concentrations, that had been prepared in the laboratory. These data were then used in order to calibrate the instrument by finding the relationship between instrument output and methanol concentration. A total of 15 standard gases were prepared by mixing a methanol gas (107 ppm) and pure air into Tedlar sample bags. Metering of gases was done by using a Tylon flow controller that was accurate within ±1%. The exponential model described by Equation (1) was employed to curve-fit the data obtained for each of the standard gases. This nonlinear model was used for calibration through the study to measure methanol levels in samples.

$$R = \alpha(1 - e^{-\beta C_s}) \tag{1}$$

where the symbols are defined as follows.

C_s: concentration of methanol in sample cell (ppm)

R: analyzer response (millivolts)

α, β: parameter of model described by Equation (1)

With measured values for C_S and R, parameters α and β were estimated by using the Gauss-Newton nonlinear least square method. Obviously, the calibration curve depends on the methanol concentration in the GFC. Consequently, calibration was conducted periodically and whenever the methanol gas in the GFC was replaced. The model, Equation (1), fit the calibration data very closely with multiple correlation coefficients greater than 0.99 in all cases.

The GFC methanol concentration was varied to find an optimum condition based on the analyzer sensitivity and degree of interference of gases existing in automobile exhausts. Five different methanol concentrations were tested ranging from 15.3 torr to 122 torr. During these tests, analyzer response was measured for a constant methanol concentration of 107 ppm in the sample cell. A plot of analyzer response versus partial pressure of methanol in the GFC is shown in Figure 3. It is apparent that the response signal approaches an asymptote that limits the maximum response for the GFC methanol pressure in excess of 122 Torr. The Interference tests were conducted to determine whether compounds normally present in automobile exhaust would cause the analyzer to respond. Diluted exhaust from a gasoline-powered vehicle was used for these checks since the exhaust contained no detectable level of methanol. Typical compounds and their concentrations existing in the automobile exhaust are given in Table 1. Results indicated barely perceptible, negative responses at each of the different GFC concentrations. Thus the instrument ability to reject interference appeared to be fairly independent of the GFC concentrations examined. The interfering gases with uncorrelated spectral absorption lines to those of methanol reduce the intensity of the filtered beam slightly more than the intensity of the unfiltered beam. For this reason, negative interferences occurred as anticipated during the tests. The largest interference observed was a negative response of 3 ppm during a special interference test. This test was run with a GFC methanol pressure of 122 torr, using a sample that contained a high hydrocarbon concentration of 178 ppm.

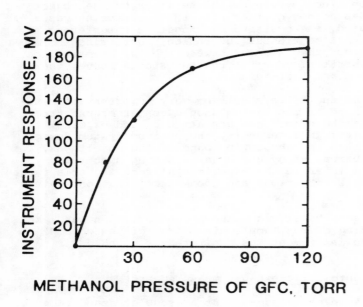

Figure 3. Change of instrument response for various GFC concentrations.

Figure 4. Response curve of methanol analyzer.

Table 1. Concentration of Interfering Gases in Automobile Exhausts

Component	Concentration
Total HC's	31 ppm
CO	260 ppm
NO_x	10 ppm
CO_2	0.5%
Formaldehyde	0.1 ppm

Table 2. Measurement Results of Low Concentration Methanol in Automobile Exhausts and Comparison with GC Results

GFCS (ppm)	GC (ppm)
0.7	0.6
0.9	0.7
0.9	0.7
1.2	0.8

Other analyzer characteristics examined were analyzer response time, signal drift, precision, and the overall detection limit. During these experiments, the partial pressure of methanol in the GFC was held at the normal condition of 122 torr, which is considered to be the optimum value for the current design.

The analyzer response time was measured while injecting a sample that generated a nearly full-scale response of the instrument. Figure 4 shows the actual trace of the instrument response obtained on a strip chart. The time indicated on the chart was measured from the point when the sample was injected. About 75 seconds were needed for the signal to reach the full response. The sample flow rate to the cell was about 15 ℓ/min. Even though a fast response was feasible by reducing the sample cell volume or increasing the sample-gas flow rate, no attempt was made at this point in the development to decrease the response time.

The signal drift of the instrument that was built on an optical table, though not serious, appeared to be the most prominent problem. The variation in surrounding temperature was found to be largely responsible for the drift. The drift occurred mostly when the GFC became inadvertently cooled during liquid nitrogen replenishment of the detector dewar. Even so, the analyzer zero drift was measured at less than 0.5 ppm over a period of 10 minutes that was sufficient enough to perform a typical analysis. The drift could easily have been reduced substantially, if the instrument had been well insulated from surrounding temperature fluctuations.

Analyzer precision was examined by running a series of measurements on three different standard gases of known concentrations. Five measurements were performed on each gas. The instrument precision, expressed as the relative standard deviation of the measurements, was 5.5%, 1.2% and 1.0% for the methanol concentrations of 7 ppm, 38 ppm and 113 ppm, respectively.

Instrument overall detection limit can be interpreted as the smallest signal that the instrument could detect reliably under normal operation conditions. The detection limit of the instrument was examined by observing instrument responses to low-level methanol concentrations with background noises and other interfering gases present in automobile exhausts. In the tests, the exhausts were diluted by mixing them with the ambient air, as shown in Figure 5, to simulate the conditions of interfering gases in real-road environments. The methanol concentrations in the sample gases were controlled by continuously injecting a small predetermined amount of methanol into the exhausts of a vehicle using only gasoline as a fuel. Reliable signal responses were obtained at methanol concentrations less than 0.5 ppm. The statistical limit of detection of an analyzer can be determined as three times the standard deviation of the baseline noise[14]. Based on this criterion the statistical limit of detection was about 1 ppm. Additional tests were also conducted to compare the GFCS results with the gas chromatograph (GC) analyses. The samples for both analyzers were collected at the same time and location as indicated in the diagram. The comparison of the results is illustrated in Table 2. By considering the results from the tests for detection limit, the methanol analyzer is capable of measuring concentrations well below 1 ppm under normal operation conditions with all adverse effects.

Comparison of the methanol analyzer results with those of the GC was continued by bag-sampling exhaust gases from a vehicle powered by a methanol-gasoline fuel blend and by measuring the methanol contents in the bags. For the GC analyses, the methanol in exhaust gases was trapped in aqueous solutions in impingers as recommended in Reference 1. These two types of samples were collected at the same time and the same location as shown in Figure 5. To obtain a wide range of methanol concentrations, samples were collected by driving the car in different modes on a chassis dynamometer or by using evaporative vapors from an enclosure surrounding the car. The summary of the measurements by the two different instruments are listed in Table 3. Measurements of the lowest concentration were 0.8 ppm for the methanol analyzer and 0.7 ppm for the GC, respectively. This concentration approached the detection limits for both instruments. In general, agreement between the two methods was good. Differences were largely from the precision limitations inherent in both methods. Precision for the methanol analyzer was discussed in the previous section. As compared with this, the precision of the GC, not including the sampling procedure, was 12% and 5% for methanol gas concentrations of 7 ppm and 113 ppm.

In addition to the analyses of bag samples, a test was conducted to check the instrument capability for real-time monitoring of emissions from methanol powered automobiles. Figure 6 shows a trace of the methanol emissions versus time. This monitoring was conducted by sampling a steady stream of diluted exhaust with the methanol analyzer. Reading the trace from right to left, methanol emissions increased rapidly after the methanol car was started in a cold environment. Concentrations soared above 110 ppm before going off the full scale of the strip chart. Then the concentrations fell rapidly and moved toward baseline levels due to catalyst activation in the automobile exhaust system. The test demonstrated the instrument potential as a convenient tool for continuous measurement of automobile emissions.

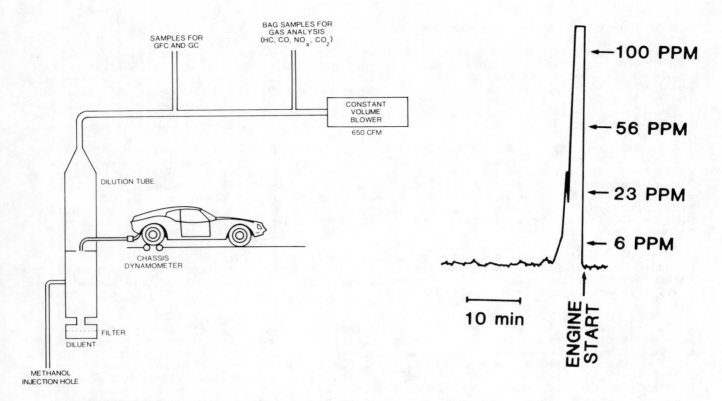

Figure 5. Schematic of sampling system of automobile emissions.

Figure 6. Real-time measurement of methanol emissions from an automobile.

Table 3.   Measurements of Methanol Concentrations in Automobile Exhausts by GFCS and GC

Driving Mode	No. of Tests	Aug. of GFCS Measurements (ppm)	Aug. of GC Measurements (ppm)
1	2	0.8	0.7
2	2	1.1	0.8
3	3	2.6	1.9
4	3	4.9	3.6
5	2	10.3	8.4
6	2	17.0	17.6
7	4	22.4	22.9
8	3	32.5	27.5
9	1	36.5	33.4
10	4	46.8	49.5

Conclusions

The instrument tested in the current study employs the correlation principle of infrared absorption lines between the gas being measured and the filter gas. The development was successful by demonstrating measurements of low concentration methanol gases in automobile exhausts. The instrument produced highly predictable outputs and consistent calibration curves. The calibration curves described by an exponential function as shown in Equation (1) resulted in very good multiple correlation coefficients. Measurements of methanol concentrations as low as 0.5 ppm were possible by using the laboratory-built instrument. The precision of the tested instrument varied from 5.5% to 1.2% for methanol concentrations ranging from 7 ppm to 113 ppm. The sensitivity and discrimination against interfering gases were adequate enough for most purposes when concentrations of methanol vapor in automobile exhausts were measured. Much of the good discrimination against non-methanol compounds that absorb in the same spectral interval is due to the correlation principle using the GFC. The sensivity of the instrument approached an upper limit as the GFC methanol concentration increased. For the design employed in this study, the methanol partial pressure of 122 torr in the GFC was adequate for maintaining the maxima sensitivity while allowing good discrimination against interfering gases. The study demonstrated good agreement between the results from the GFCS with those from the GC when methanol emissions from an automobile were measured. The other important features of the instrument are its simplicity of use and its speed of sample analysis, as compared with wet chemical methods using a GC. Therefore the instrument can be effectively used for real-time measurement of methanol concentrations. A small drift in the zero setting was primarily due to variations in the temperature of the components. Hence an improvement in stability can be obtained by thermally insulating the detector dewar containing liquid nitrogen from the other components and maintaining a constant temperature of the instrument with an enclosure. Sensitivity to methanol vapor and interference of other gases both depend on the spectral band utilized. The spectral band tested was estimated by inspection of the infrared absorption spectrum of methanol. The optimum central wavelength and band width need further investigation to improve the performance of the system. Most of the gaseous pollutants appearing in automobile exhausts can be measured by utilizing their infrared absorption. Hence the GFCS has a very strong potential to expand its application scope to measurements of other gaseous pollutants.

References

1. L.R. Smith and C. Urban, Characterization of Exhaust Emissions from Methanol and Gasoline Fueled Automobiles, Final Report EPA460/3-82-004, U.S. Environmental Protection Agency, Research Triangle Park, NC, 1982.
2. D.E. Burch, F.J. Gates, D.A. Gryvnak, and J.D. Pembrook, Versatile Gas Filter Correlation Spectrometer, Final Report, EPA 600/2-75-024, U.S. Environmental Protection Agency, Research Triangle Park, NC., 1975.
3. P.A. Gabele and S. Cha, Methanol Measurement in Auto Exhaust Using a Gas-Filter Correlation Spectrometer, Technical Paper 852137, Society of Automotive Engineers, Warrendale, PA, 1985.
4. N. Wright and L.W. Hersher, J. Opt. Soc. Am. 36(4), 195 (1946).
5. W.G. Fastie and A.H. Pfund, J. Opt. Soc. Am. 37(10, 762 (1947).
6. R.C. Fowler, Rev. Sci. Instrum. 20(3), 175 (1949).
7. D.W. Hill and T. Powell, Nondispersive Infrared Gas Analysis in Science, Medicine and Industry, Chap., 1, Plenum Press, New York (1968).
8. P. Goody, J. Opt. Soc. Am. 58(7), 900 (1968).
9. E.R. Bartle, S. Kaye and E.A. Meckstroth, J. Spac. Rock. 9(11), 836 (1972).
10. R.K. Stevens and W.F. Herget, Analytical Methods Applied to Air Pollution Measurements, Chap. 10, Ann Arbor Science Publishers, Inc., Ann Arbor (1974).
11. D.I. Sebacher, A Gas Filter Correlation Monitor for Co, CH_4 and HCl, NASA Technical Paper No. 1113, p. 28 (1977).
12. J.V. White, J. Opt. Soc. Am. 32(5), 285 (1942).
13. R.H. Pierson, A.N. Fletcher and E.C. Gantz, Anal. Chem. 28(8), 1218 (1956).
14. L.H. Keith, R.L. Libby, W. Crummett, J.K. Taylor, J. Daegan, Jr., and G. Wentler, Anal. Chem. 55(14), 2210 (1983).

Monitorization technology applied to metal cutting

J. Lipták, D. Kozáková, M. Ábel, I. Maňková

Technical University, Švermova 9, Košice 040 01, ČSSR

V. Modrák

International Robot Association, Prešov, ČSSR

M. Štovčík

East-Slovakian Iron and Steel Works, Košice, ČSSR

Abstract

Monitorization and regulation of the chip behaviour in the cutting zone present one of the major problems in the reliable operation of flexible manufacturing systems. The chips should be formed so as to eliminate the tool damage and enable their easy disposal from the cutting zone. For solving the problem a convenient regulation strategy should be developed. In the development of the strategy the authors extended the theoretical and experimental works on a special experimental device, which makes the cutting process modelling possible, and carried out a number of studies. The paper presents the philosophy of the problem solution and the results of experimental studies. The experimental system is composed of a device, which allows to observe the cutting zone, and systems for modelling and dual monitoring operations. The work includes the description of experimental subsystems and analyzes the results of the observed chip states. The authors suggest a formulation for the state control strategy and base their research on the graphical computer interpretation of the chip states. The system developed makes use of TV and IR cameras which allow to investigate the cutting zone. The advanced recording and computing technologies enable a real time recognition of the chip states and thermal fields both in the tool and the chip.

Introduction

In unmanned operations of manufacturing machines the chip shape and removal control is one of the most important problems. To avoid the problems caused by chips most of the present flexible manufacturing systems are restricted to brittle work materials or intermittent cutting operations. There are three stages of the problem caused by the chips :
1. Chip shape control at the beginning of chip formation
2. Chip collection and removal from the cutting zone to the place of processing
3. Chip waste processing.

Chip formation is one of the most difficult problems mainly in continuous cutting operations. The shape of chips changes stochastically even in fixed cutting conditions which is caused by the variable bending moment in the chip root. The bending moment is effected by the increasing chip mass and the obstructions in the chip flow. The chip shape is largely determined if all conditions of influence are fixed and stable. Geometrically it is given by the combination of several characteristics with the chip thickness, width and length. To understand and regulate the chip shape mechanism some more fundamental studies will be necessary.

Automatic chip observation in specific situations and adaptive control or stop of the process in the moment when chips begin to cause problems may reduce the risk of failures and stop the manufacture. Therefore sufficient and reliable control of the chip shape in the process of its formation is essential for automation of manufacturing operations. To realize this task it is important to use an appropriate measuring quantity and work out the control strategy. Great complexity of the chip formation process makes it necessary to apply complicated monitorization systems which enable to identify the chip formation mechanism and regulate the cutting process (by the adaptive control).

Laboratory concept of the research

The investigation strategy and its successful realization under the conditions of flexible manufacturing systems is led and coordinated by prof. Buda.

The system presented is a dual monitorization system based on telerecorder and thermovision technology integrated with mechanical cutting modules. The integrated system (Fig. 1) is based on the module of cyclic cut interrupter. The fundamental principle of the cutting module lies in orthogonal cutting of the tool in the specimen of defined shape and dimensions. The main feature of the module structure is the unit for rigid clamping of the specimen which allows to adjust the specimen position and provide the feed in the cutting direction. The module position is fixed in the process of cutting. Coordinate tool setting and constant cutting rate are also possible. The construction design places the cutting module into a frame which provides for the physical integration of the regulated drive and the monitorization system. The monitorization system consists of a combination of TV and IR cameras and makes possible a lateral observation of the cutting zone and the chip which is being formed. By means of a single-channel TV circuit of the camera and the monitor it is possible to obtain video records of the process. Fig. 2 is an illustration of the arrangement and movable assembly of the TV and IR cameras with respect to the cut interrupter. Both cameras are opposite to each other. In a focal length between them there are the specimen (20x20x3) and the tool whose interaction represents the cutting zone. The laboratory monitorization system UM CPR 16 is shown in Fig. 3. The cameras work on different principles independently on each other. The thermovision camera AGA 782 with the complete equipment, computer and printer, has been used in interface with the TV camera of the CCD type (Hitashi). Video records are obtained in two ways - either by the use of the built-in adapter circuit to which a modified VTR (video tape recorder) is connected, or through the optional computer equipment DISCON (digital infrared system for colouration). The former allows to use the switch PICTURE MODE to select the type of picture and the ISOTHERM LEVEL control to analyze the displayed picture of the thermovision camera according to the operator's demands. By means of the latter it is possible to connect a standard VCR (video cassette recorder) directly to the DISCON. When using the first of the two methods the picture may be played in the same form only in which it was recorded. DISCON has the advantage of using standard video equipment both for recording and playing. However it is also possible to use both methods independently and transmit the pictures recorded by VTR to VCR. The IR camera output is stored in disks and when necessary for analysis it may be interpreted on the plotter or the display in a graphic representation.

Laboratory studies that were carried out showed the possibility of the cutting zone monitorization by the systems of both cameras. The analysis of the records obtained for formulating the chip control strategy shows that the results of conventional research studies may be correlated with the outputs of the cutting zone monitorization in the form of recorded pictures. Correlation calculations prove the validity of the influence of technological conditions on the formation and development of chips stated by means of earlier methods.

Process of modelling on the lathe and control strategy

The results of the experiments carried out on the laboratory equipment UM CPR 16 were used in the NC lathe workshop. Observation and monitorization of the cutting process with the use of the TV camera is shown in Fig. 4 and 5. When taking pictures in the horizontal level the camera is fixed so that the optical line lies in the plane Ps of the cutting tool edge. The field of view of the camera is focused to a part of the tool and the work surface as well as the chip removal direction. The optical length between the objective and the camera optics is about 30 cm. When locating the camera in the vertical level the optical line of the camera in the plane Po is perpendicular to the main cutting tool edge. The field of view and the optical length are the same as in the former case. Fig. 6 shows some records obtained by means of photographic reproduction from the CCD camera monitor. The image sharpness may often be small which is caused by the low image frequency of the camera. The cameras with higher image frequency provide better results. The CCD camera proved to be useful for studies carried out on the laboratory equipment, but in commercial lathe applications it is connected with great difficulties concerning the image digitalization for the control strategy. The image obtained by this type of the camera must be relieved from the chip environment. The qualitative evaluation and quantification of image differences of the CCD cameras must be done by the human operator.

The image digitalization problem may be solved by the thermovision camera with computer processing of the process observed. Fig. 7 shows the scheme of the AGA camera application in face turning. The illustration of the system outputs is shown in Fig. 8. The first two frames show a segmented shape characteristic for the normal cutting process. The third frame illustrates the controllable shape which signalizes the approach to difficult to control states (the fourth frame). The critical shape (the fifth frame) points to dangerous states (the sixth frame) with a highly probable tool failure which should be eliminated from the flexible manufacturing systems. The thermovision pictures obtained from the thermal field records make it possible to correlate the areas of characteristic temperatures to those of the total frame area thus gaining the quantifiable and measurable parameter from which the control strategy for the chip states in the cutting zone may be derived. On the basis of the logic development of this parameter the necessary software may be designed.

The thermovision pictures are relatively simple and usually no problems arise in their fragmentation. In the time relations of the machining process the temperature dependence on the tool wear represents the higher order polynomial whose interpretation makes it necessary to apply the spline approximations which may be used for the concrete manufacturing system control. The related problems however are the subject of our present research into the indication and prediction possibilities for the cutting process failures.

Micro-computer control system

The micro-computer feedback control system was designed in connection to the servo motion (Fig. 9). The basic unit of the micro-computer control system is the central processor with the Z 80 CPU micro-processor. This micro-processor satisfies the necessary flexibility and speed of the system which is given mainly by the set of instructions. The most important part of the system are the input/output circuits whose errors may introduce inaccuracy into the system functions. The output circuits control the servo drives, therefore it is necessary to modify the output signal to the given type of servo systems. In our case the system controls the chuck revolutions, the cross and longitudinal feed of the carriage and the switch on/off state. The system uses three physical inputs :
1. Piezo-resistive compression sensor to follow the acoustic emission and the loading force of the cutting tool. The two functions are multiplexed in the processor frequency.
2. The input from the camera scanning the chip shape and amount.
3. The input from the thermovision camera of the AGA system which is only a temporary solution. The inherent thermal input is being planned on the basis of special thermocouples. The thermal analysis is decoded by the convertor controlled micro-processor.
The information obtained by sensors is evaluated by the program on the basis of comparison with the predetermined nominal constants and the feedback control makes corrections of the equipment functions. For example, according to the chip shape and temperature it is possible to find the level of chip segmentation and cutting tool wear. The system can distinguish four states of the cutting tool wear:
1. Nominal value - no corrections are made.
2. Slight corrections of the cutting conditions are necessary.
3. Substantial change of the cutting conditions.
4. The equipment is out of operation, the tool is replaced.

The control process is executed in the real time. An essential part of the system is the so-called journal mechanism which writes continuously all inputs and outputs into the magnetic medium. In the case of failure the process may be simulated and the failure causes analyzed. The journal mechanism is also applicable to some other purposes - testing the new modules, etc. The micro-computer contrl system is universal. Its software modification may extend the application possibilities. Further subsystems intefaced with the local network may increase the potential of the controlled channels.

Conclusions

The long-term development in the field of research into identification of the phenomena in the cutting zone led at our workplace to the construction of the universal monitorization system prototype which can provide information from the cutting zone and represent it in the form of quantified (digital and binary) data. The knowledge obtained was adapted to the work conditions of NC lathes. The results achieved are presentations formulated on the level of outputs from model experiments. They reflect the synthesis of process regulation with the mechanics of the built equipment and the information technology. The application of the infrared scanning technology proved to be useful for the purpose of chip state quantifications. This approach stresses the principles of distinguishing the temperature states in the tool and the chip, thus regenerating the temperature re - search in the cutting zone. The infrared distinction technology may reliably be applied to the cutting zone monitorization and the state parameters obtained may be used for distinction and control strategy design as well as for hardware, firmware and software design in the control of work stations, flexible manufacturing cells and flexible manufacturing systems.

References

1. Arai, M.- Nakayama, K.- Tsukada, Y.: The monitoring of chip flow in metal cutting using radiation thermometer, Proceedings of the 5th international conference on production engineerging, Tokyo, 1984
2. Buda, J.- Lipták, J.- Šefara, M.: A new position research and evaluation of metal machinability, Technical paper, Westec 80, Society of manufacturing engineers, Dearborn, Michigan, NR 80-221
3. Buda, J.- Lipták, J.: Characteristics and consequences of motional domains in the cutting zone, Proceedings of the 5th international conference on production engineering, Japan society of precision engineering, Tokyo, 1984
4. Buda, J.- Lipták, J.- Modrák, V.- Ábel, M.- Stovčík, M.- Maňková, I.: Chip state monitorization in flexible manufacturing systems, Proceedings of the 18th CIRP MFS-S, Stuttgart, F.R.Germany, 1986

5. König, W.- Luttervelt, C.A.- Nakayama, K.: Present knowledge of chip control, CIRP Annals 28, 1979, 4
6. Lipták, J.- Kažimír, I.- Gazárek, G.: Mikrokinematografičeskij metod izučenija dviženija elementov v strukture obrabatimajemovo metala, Proceedings of the 14th international congress on high speed photography and photonics, Moscow, 1980
7. Lipták, J.: Utilizing high speed film in metal cutting research, Proceedings of the 15th international congress on high speed photography and photonics, The international society for optical engineering, San Diego, California, USA, 1982, No. 348-08
8. Lipták, J.: New ways of identifying the continuous process occurring in metal cutting, Proceedings of the 5th ICREC, International Colloquium of research and education cinematography, Brno, ČSSR, 1983.

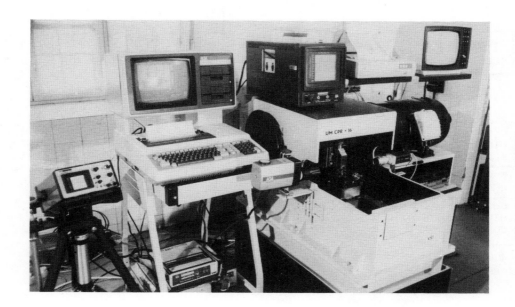

Figure 1. Laboratory monitorization system UM CPR - 16.

Figure 2. TV and thermocamera arrangement in the cyclic cut interrupter.

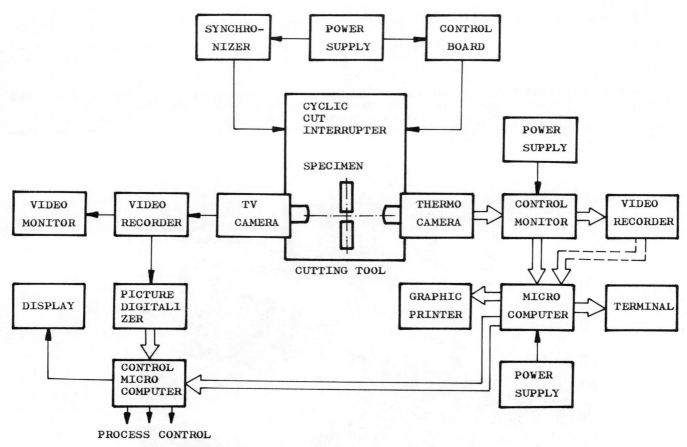

Figure 3. Monitoring system model

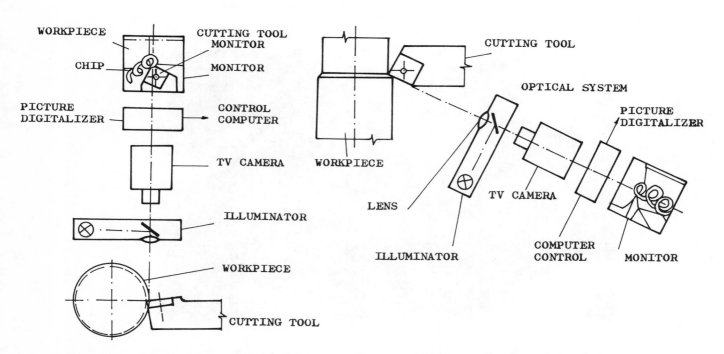

Figure 4. Chip shape monitorization in the plane P_o

Figure 5. Chip shape monitorization in the plane P_s

Figure 6a View from the plane P_s

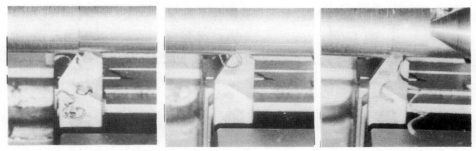

Figure 6b View from the plane P_o

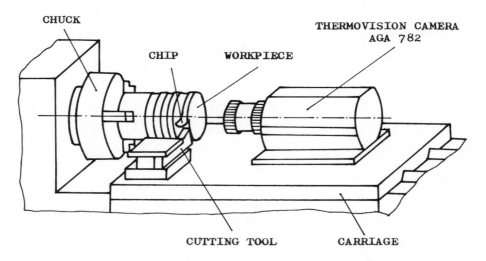

Figure 7. AGA camera installation in face turning

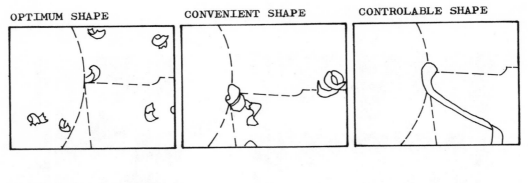

Figure 8. Comparison thermovision graphic records of the chip shapes in turning

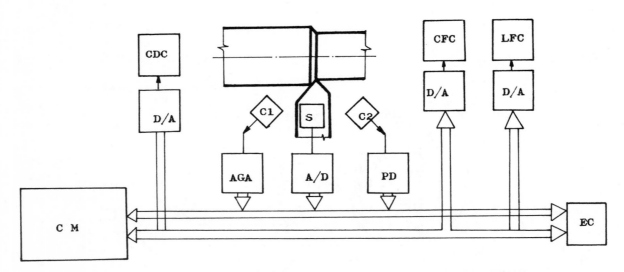

Figure 9. Functional principle of the cutting process monitorization

D/A,A/D - D/A,A/D convertor

C1 AGA - thermovision camera using AGA system,

S - piezoresistive compresion sensor

CFC - cross feed control of the carriage

EC - extension for additional control and scanning components

PD - picture digitalizer

CDC - chick drive control

C2 - chip shape scanning camera

LFC - longitudal feed control of the carriage

CM - control microcomputer

INFRARED TECHNOLOGY XII

Volume 685

Addendum

The following papers, which were scheduled to be presented at this conference and published in these proceedings, were cancelled.

[685-09] **Ballistic missile IR signature modeling (Invited Paper)**
N. Aldrich, B. Klem, The Aerospace Corp.

[685-29] **Performance of an IR spatial radiometer (Invited Paper)**
G. P. Wilson, R. P. Reinker, Ball Aerospace Systems Div.; D. Paulsen, Air Force Geophysics Lab.

AUTHOR INDEX